Essentials of Oceanography

Essentials of Oceanography

Edited by
Theobald Lane

Larsen & Keller
www.larsen-keller.com

Essentials of Oceanography
Edited by Theobald Lane
ISBN: 978-1-63549-203-3 (Hardback)

Larsen & Keller

Published by Larsen and Keller Education,
5 Penn Plaza,
19th Floor,
New York, NY 10001, USA

Cataloging-in-Publication Data

Essentials of oceanography / edited by Theobald Lane.
   p. cm.
Includes bibliographical references and index.
ISBN 978-1-63549-203-3
1. Oceanography. 2. Ocean. 3. Marine sciences. 4. Marine pollution. I. Lane, Theobald.
GC11.2 .E87 2017
551.46--dc23

The publisher's policy is to use permanent paper from mills that operate a sustainable forestry policy. Furthermore, the publisher ensures that the text paper and cover boards used have met acceptable environmental accreditation standards.

Printed and bound in the United States of America.

For more information regarding Larsen and Keller Education and its products, please visit the publisher's website www.larsen-keller.com

Table of Contents

Permissions

Index

Preface

Oceanology or oceanography is a sub-field of Earth sciences. It is concerned with the study of oceans. It includes topics like waves, ecosystem dynamics, seafloor geology, plate tectonics, physical properties and geophysical fluid dynamics, etc. This book provides thorough insights into the field of oceanography. The topics included in it on the subject are of utmost significance and are bound to provide incredible insights to readers. It also explores all the important aspects of this area in the present day scenario. Coherent flow of topics, student-friendly language and extensive use of examples make this book an invaluable source of knowledge.

To facilitate a deeper understanding of the contents of this book a short introduction of every chapter is written below:

Chapter 1- The branch of earth science that deals with the ocean is referred to as oceanography. It covers a wide range of topics which includes ocean currents, waves and all the complex features of the ocean. This chapter is an overview on the ecology of oceans and provides with an integrated understanding of oceanography.

Chapter 2- Oceanography has many branches as a subject. This chapter will provide a glimpse of related fields of oceanography. Physical oceanography is one of the several branches of oceanography; the other branches include chemical oceanography, biological oceanography, acoustical oceanography and paleoceanography.

Chapter 3- Measuring large portions of the ocean for a better understanding of the temperature of the ocean is known as ocean acoustic tomography while the study of the depth of the lake and ocean floor is bathymetry. The chapter serves as a source to understand the major categories related to oceanography, providing the reader with a detailed insight on the methods used in oceanography.

Chapter 4- An ocean is a body of water, which has significant amounts of dissolved salt in it. Some of the oceans discussed in this chapter are the Artic Ocean, Atlantic Ocean, Pacific and Indian Ocean. It provides examples of and also gives a brief introduction on the deepest part of the world ocean. This chapter is an overview of the subject matter incorporating all the major aspects of ocean.

Chapter 5- Ocean can be best understood in confluence with the major topics listed in the following chapter. Oceans cover 71 % of the surface of earth. Some of the aspects are ocean current, ocean heat content and plate tectonics. This chapter helps the reader to broaden the existing knowledge on oceans.

Chapter 6- Ocean reanalysis is a method of combining historical ocean observations with a general ocean model. In recent time, a number of efforts have been put in to estimate the physical state of the ocean. This chapter provides the reader with a detailed insight on ocean observations and on the reanalysis of oceans.

Chapter 7- The impact of pollution on oceans has drastically increased. Eighty percent of marine pollution comes from land. Ocean population is as important as land pollution, and measures need to be taken to prevent it. The chapter serves as a source to understand marine pollution, and helps the reader to develop a deep understanding on the subject.

I owe the completion of this book to the never-ending support of my family, who supported me throughout the project.

Editor

Introduction to Oceanography

The branch of earth science that deals with the ocean is referred to as oceanography. It covers a wide range of topics which includes ocean currents, waves and all the complex features of the ocean. This chapter is an overview on the ecology of oceans and provides with an integrated understanding of oceanography.

Oceanography, also known as oceanology, is the branch of Earth science that studies the ocean. It covers a wide range of topics, including ecosystem dynamics; ocean currents, waves, and geophysical fluid dynamics; plate tectonics and the geology of the sea floor; and fluxes of various chemical substances and physical properties within the ocean and across its boundaries. These diverse topics reflect multiple disciplines that oceanographers blend to further knowledge of the world ocean and understanding of processes within: astronomy, biology, chemistry, climatology, geography, geology, hydrology, meteorology and physics. Paleoceanography studies the history of the oceans in the geologic past.

History

Map of the Gulf Stream by Benjamin Franklin, 1769-1770. Courtesy of the NOAA Photo Library.

Early History

Humans first acquired knowledge of the waves and currents of the seas and oceans in pre-historic times. Observations on tides were recorded by Aristotle and Strabo. Early exploration of the oceans was primarily for cartography and mainly limited to its surfaces and of the animals that fishermen brought up in nets, though depth soundings by lead line were taken.

Although Juan Ponce de León in 1513 first identified the Gulf Stream, and the current was well-known to mariners, Benjamin Franklin made the first scientific study of it and gave it its name. Franklin measured water temperatures during several Atlantic crossings and correctly explained the Gulf Stream's cause. Franklin and Timothy Folger printed the first map of the Gulf Stream in 1769-1770.

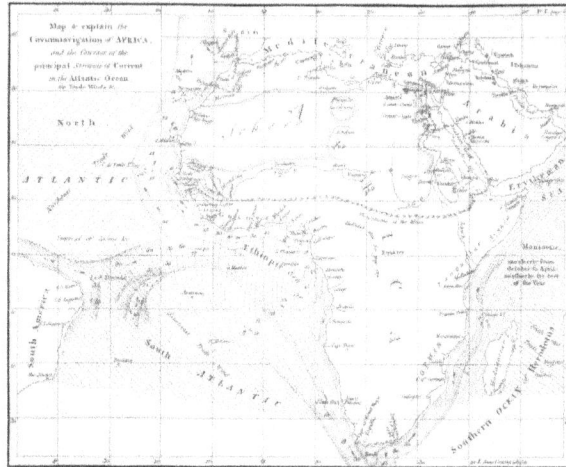

1799 map of the currents in the Atlantic and Indian Oceans, by James Rennell

Information on the currents of the Pacific Ocean was gathered by explorers of the late 18th century, including James Cook and Louis Antoine de Bougainville. James Rennell wrote the first scientific textbooks on oceanography, detailing the current flows of the Atlantic and Indian oceans. During a voyage around the Cape of Good Hope in 1777, he mapped *"the banks and currents at the Lagullas"*. He was also the first to understand the nature of the intermittent current near the Isles of Scilly, (now known as Rennell's Current).

Sir James Clark Ross took the first modern sounding in deep sea in 1840, and Charles Darwin published a paper on reefs and the formation of atolls as a result of the Second voyage of HMS Beagle in 1831-6. Robert FitzRoy published a four-volume report of the Beagle's three voyages. In 1841–1842 Edward Forbes undertook dredging in the Aegean Sea that founded marine ecology.

The first superintendent of the United States Naval Observatory (1842–1861), Matthew Fontaine Maury devoted his time to the study of marine meteorology, navigation, and charting prevailing winds and currents. His 1855 textbook *Physical Geography of the Sea* was one of the first comprehensive oceanography studies. Many nations sent oceanographic observations to Maury at the Naval Observatory, where he and his colleagues evaluated the information and distributed the results worldwide.

Modern Oceanography

Despite all this, human knowledge of the oceans remained confined to the topmost few fathoms of the water and a small amount of the bottom, mainly in shallow areas. Almost nothing was known of the ocean depths. The Royal Navy's efforts to chart all of the world's coastlines in the mid-19th century reinforced the vague idea that most of the ocean was very deep, although little more was known. As exploration ignited both popular and scientific interest in the polar regions and Africa, so too did the mysteries of the unexplored oceans.

H.M.S. CHALLENGER UNDER SAIL, 1874.

HMS *Challenger* undertook the first global marine research expedition in 1872.

The seminal event in the founding of the modern science of oceanography was the 1872-76 Challenger expedition. As the first true oceanographic cruise, this expedition laid the groundwork for an entire academic and research discipline. In response to a recommendation from the Royal Society, The British Government announced in 1871 an expedition to explore world's oceans and conduct appropriate scientific investigation. Charles Wyville Thompson and Sir John Murray launched the Challenger expedition. The Challenger, leased from the Royal Navy, was modified for scientific work and equipped with separate laboratories for natural history and chemistry. Under the scientific supervision of Thomson, Challenger travelled nearly 70,000 nautical miles (130,000 km) surveying and exploring. On her journey circumnavigating the globe, 492 deep sea soundings, 133 bottom dredges, 151 open water trawls and 263 serial water temperature observations were taken. Around 4,700 new species of marine life were discovered. The result was the *Report Of The Scientific Results of the Exploring Voyage of H.M.S. Challenger during the years 1873-76*. Murray, who supervised the publication, described the report as "the greatest advance in the knowledge of our planet since the celebrated discoveries of the fifteenth and sixteenth centuries". He went on to found the academic discipline of oceanography at the University of Edinburgh, which remained the centre for oceanographic research well into the 20th century. Murray was the first to study marine trenches and in particular the Mid-Atlantic Ridge, and map the sedimentary deposits in the oceans. He tried to map out the world's ocean currents based on salinity and temperature observations, and was the first to correctly understand the nature of coral reef development.

In the late 19th century, other Western nations also sent out scientific expeditions (as did private individuals and institutions). The first purpose built oceanographic ship, the Albatros, was built in 1882. In 1893, Fridtjof Nansen allowed his ship, Fram, to be frozen in the Arctic ice. This enabled him to obtain oceanographic, meteorological and astronomical data at a stationary spot over an extended period.

Between 1907 and 1911 Otto Krümmel published the *Handbuch der Ozeanographie*, which became influential in awakening public interest in oceanography. The four-month 1910 North Atlantic expedition headed by John Murray and Johan Hjort was the most ambitious research oceanographic and marine zoological project ever mounted until then, and led to the classic 1912 book *The Depths of the Ocean*.

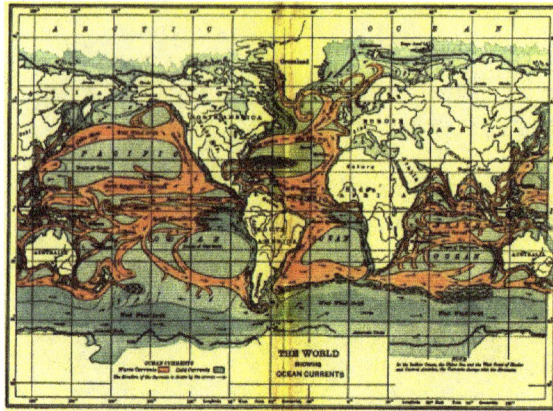

Ocean currents (1911)

The first acoustic measurement of sea depth was made in 1914. Between 1925 and 1927 the "Meteor" expedition gathered 70,000 ocean depth measurements using an echo sounder, surveying the Mid-Atlantic ridge.

Sverdrup, Johnson and Fleming published *The Oceans* in 1942, which was a major landmark. *The Sea* (in three volumes, covering physical oceanography, seawater and geology) edited by M.N. Hill was published in 1962, while Rhodes Fairbridge's *Encyclopedia of Oceanography* was published in 1966.

The Great Global Rift, running along the Mid Atlantic Ridge, was discovered by Maurice Ewing and Bruce Heezen in 1953; in 1954 a mountain range under the Arctic Ocean was found by the Arctic Institute of the USSR. The theory of seafloor spreading was developed in 1960 by Harry Hammond Hess. The Ocean Drilling Program started in 1966. Deep sea vents were discovered in 1977 by John Corlis and Robert Ballard in the submersible DSV *Alvin*.

In the 1950s, Auguste Piccard invented the bathyscaphe and used the *Trieste* to investigate the ocean's depths. The United States nuclear submarine Nautilus made the first journey under the ice to the North Pole in 1958. In 1962 the FLIP (Floating Instrument Platform), a 355-foot spar buoy, was first deployed.

From the 1970s, there has been much emphasis on the application of large scale computers to oceanography to allow numerical predictions of ocean conditions and as a part of overall environmental change prediction. An oceanographic buoy array was established in the Pacific to allow prediction of El Niño events.

1990 saw the start of the World Ocean Circulation Experiment (WOCE) which continued until 2002. Geosat seafloor mapping data became available in 1995.

In recent years studies advanced particular knowledge on ocean acidification, ocean heat content, ocean currents, the El Niño phenomenon, mapping of methane hydrate deposits, the carbon cycle, coastal erosion, weathering and climate feedbacks in regards to climate change interactions.

Study of the oceans is linked to understanding global climate changes, potential global warming and related biosphere concerns. The atmosphere and ocean are linked because of evaporation and precipitation as well as thermal flux (and solar insolation). Wind stress is a major driver of ocean currents while the ocean is a sink for atmospheric carbon dioxide. All these factors relate to the ocean's biogeochemical setup.

Branches

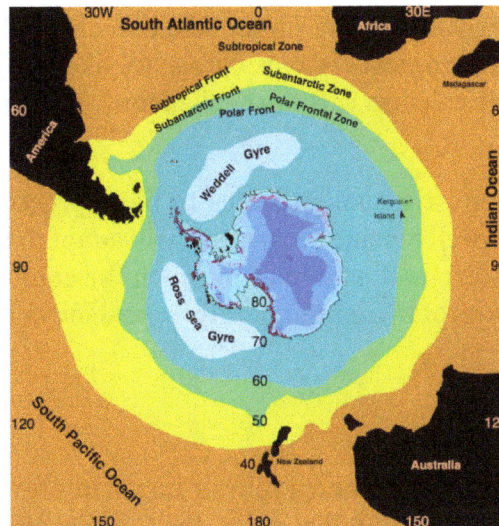

Oceanographic frontal systems on the Southern Hemisphere

The study of oceanography is divided into these four branches:

- Biological oceanography, or marine biology, investigates the ecology of marine organisms in the context of the physical, chemical, and geological characteristics of their ocean environment and the biology of individual marine organisms.

- Chemical oceanography, or marine chemistry, is the study of the chemistry of the ocean and its chemica constraint programming is a programming paradigm wherein relations between variables are stated in the form of constraints. l interaction with the atmosphere.

- Geological oceanography, or marine geology, is the study of the geology of the ocean floor including plate tectonics and paleoceanography.

- Physical oceanography, or marine physics, studies the ocean's physical attributes including temperature-salinity structure, mixing, surface waves, internal waves, surface tides, internal tides, and currents.

Ocean Acidification

Ocean acidification describes the decrease in ocean pH that is caused by anthropogenic carbon dioxide (CO_2) emissions into the atmosphere. Seawater is slightly alkaline and had a preindustrial pH of about 8.2. More recently, anthropogenic activities have steadily increased the carbon dioxide content of the atmosphere; about 30–40% of the added CO_2 is absorbed by the oceans, forming carbonic acid and lowering the pH (now below 8.1) through ocean acidification. The pH is expected to reach 7.7 by the year 2100.

An important element for the skeletons of marine animals is calcium, but calcium carbonate becomes more soluble with pressure, so carbonate shells and skeletons dissolve below the carbonate compensation depth. Calcium carbonate becomes more soluble at lower pH, so ocean acidification is likely to affect marine organisms with calcareous shells, such as oysters, clams, sea urchins and corals, and the carbonate compensation depth will rise closer to the sea surface. Affected planktonic organisms will include pteropods, coccolithophorids and foraminifera, all important in the food chain. In tropical regions, corals are likely to be severely affected as they become less able to build their calcium carbonate skeletons, in turn adversely impacting other reef dwellers.

The current rate of ocean chemistry change seems to be unprecedented in Earth's geological history, making it unclear how well marine ecosystems will adapt to the shifting conditions of the near future. Of particular concern is the manner in which the combination of acidification with the expected additional stressors of higher temperatures and lower oxygen levels will impact the seas.

Ocean Currents

Since the early ocean expeditions in oceanography, a major interest was the study of the ocean currents and temperature measurements. The tides, the Coriolis effect, changes in direction and strength of wind, salinity and temperature are the main factors determining ocean currents. The thermohaline circulation (THC) *thermo-* referring to temperature and *-haline* referring to salt content connects 4 of 5 ocean basins and is primarily dependent on the density of sea water. Ocean currents such as the Gulf Stream are wind-driven surface currents.

Ocean Heat Content

Oceans of Climate Change NASA

Oceanic heat content (OHC) refers to the heat stored in the ocean. The changes in the ocean heat play an important role in sea level rise, because of thermal expansion. Ocean warming accounts for 90% of the energy accumulation from global warming between 1971 and 2010.

Oceanographic Institutions

The first international organization of oceanography was created in 1902 as the International

Council for the Exploration of the Sea. In 1903 the Scripps Institution of Oceanography was founded, followed by Woods Hole Oceanographic Institution in 1930, Virginia Institute of Marine Science in 1938, and later the Lamont-Doherty Earth Observatory at Columbia University, and the School of Oceanography at University of Washington. In Britain, the National Oceanography Centre (an institute of the Natural Environment Research Council) is the successor to the UK's Institute of Oceanographic Sciences. In Australia, CSIRO Marine and Atmospheric Research (CMAR), is a leading centre. In 1921 the International Hydrographic Bureau (IHB) was formed in Monaco.

Oceanographic Museum

References

- Rice, A. L. (1999). "The Challenger Expedition". Understanding the Oceans: Marine Science in the Wake of HMS Challenger. Routledge. pp. 27–48. ISBN 978-1-85728-705-9.

- "Sir John Murray (1841-1914) - Founder Of Modern Oceanography". Science and Engineering at The University of Edinburgh. Retrieved 7 November 2013.

- Gattuso, J.-P.; Hansson, L. (15 September 2011). Ocean Acidification. Oxford University Press. ISBN 978-0-19-959109-1. OCLC 730413873.

- "Ocean acidification". Department of Sustainability, Environment, Water, Population & Communities: Australian Antarctic Division. 28 September 2007. Retrieved 17 April 2013.

- Cohen, A.; Holcomb, M. (2009). "Why Corals Care About Ocean Acidification: Uncovering the Mechanism" (PDF). Oceanography. 24 (4): 118–127. doi:10.5670/oceanog.2009.102.

- Hamblin, Jacob Darwin (2005) Oceanographers and the Cold War: Disciples of Marine Science. University of Washington Press. ISBN 978-0-295-98482-7

- Steele, J., K. Turekian and S. Thorpe. (2001). Encyclopedia of Ocean Sciences. San Diego: Academic Press. (6 vols.) ISBN 0-12-227430-X

- Boling Guo, Daiwen Huang. Infinite-Dimensional Dynamical Systems in Atmospheric and Oceanic Science, 2014, World Scientific Publishing, ISBN 978-981-4590-37-2.

Branches of Oceanography

Oceanography has many branches as a subject. This chapter will provide a glimpse of related fields of oceanography. Physical oceanography is one of the several branches of oceanography; the other branches include chemical oceanography, biological oceanography, acoustical oceanography and paleoceanography.

Physical Oceanography

Physical oceanography is the study of physical conditions and physical processes within the ocean, especially the motions and physical properties of ocean waters.

Physical oceanography is one of several sub-domains into which oceanography is divided. Others include biological, chemical and geological oceanographies.

Physical Setting

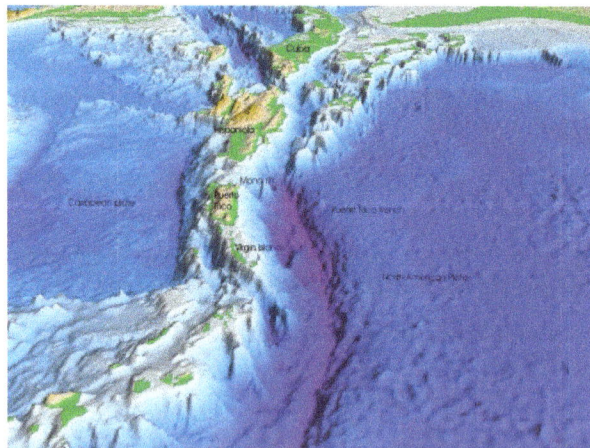

Perspective view of the sea floor of the Atlantic Ocean and the Caribbean Sea.
The purple sea floor at the center of the view is the Puerto Rico Trench.

The pioneering oceanographer Matthew Maury said in 1855 *"Our planet is invested with two great oceans; one visible, the other invisible; one underfoot, the other overhead; one entirely envelopes it, the other covers about two thirds of its surface."* The fundamental role of the oceans in shaping Earth is acknowledged by ecologists, geologists, meteorologists, climatologists, geographers and others interested in the physical world. An Earth without oceans would truly be unrecognizable.

Roughly 97% of the planet's water is in its oceans, and the oceans are the source of the vast majority of water vapor that condenses in the atmosphere and falls as rain or snow on the continents. The tremendous heat capacity of the oceans moderates the planet's climate, and its absorption of various gases affects the composition of the atmosphere. The ocean's influence extends even to the composition of volcanic rocks through seafloor metamorphism, as well as to that of volcanic gases and magmas created at subduction zones.

The oceans are far deeper than the continents are tall; examination of the Earth's hypsographic curve shows that the average elevation of Earth's landmasses is only 840 metres (2,760 ft), while the ocean's average depth is 3,800 metres (12,500 ft). Though this apparent discrepancy is great, for both land and sea, the respective extremes such as mountains and trenches are rare.

Area, volume plus mean and maximum depths of oceans (excluding adjacent seas)				
Body	**Area** $(10^6 km^2)$	**Volume** $(10^6 km^3)$	**Mean depth** (m)	**Maximum** (m)
Pacific Ocean	165.2	707.6	4282	-11033
Atlantic Ocean	82.4	323.6	3926	-8605
Indian Ocean	73.4	291.0	3963	-8047
Southern Ocean	20.3			-7235
Arctic Ocean	14.1		1038	
Caribbean Sea	2.8			-7686

Temperature, Salinity and Density

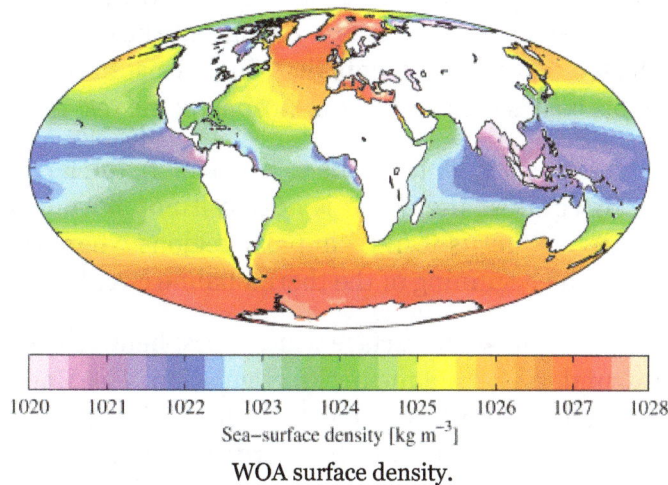

WOA surface density.

Because the vast majority of the world ocean's volume is deep water, the mean temperature of seawater is low; roughly 75% of the ocean's volume has a temperature from $0° - 5 °C$ (Pinet 1996). The same percentage falls in a salinity range between 34–35 ppt (3.4–3.5%) (Pinet 1996). There is still quite a bit of variation, however. Surface temperatures can range from below freezing near the poles to 35 °C in restricted tropical seas, while salinity can vary from 10 to 41 ppt (1.0–4.1%).

The vertical structure of the temperature can be divided into three basic layers, a surface mixed layer, where gradients are low, a thermocline where gradients are high, and a poorly stratified abyss.

In terms of temperature, the ocean's layers are highly latitude-dependent; the thermocline is pronounced in the tropics, but nonexistent in polar waters (Marshak 2001). The halocline usually lies near the surface, where evaporation raises salinity in the tropics, or meltwater dilutes it in polar regions. These variations of salinity and temperature with depth change the density of the seawater, creating the pycnocline.

Circulation

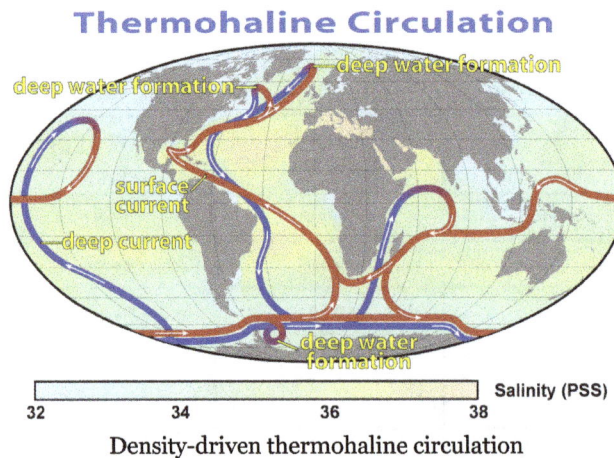

Density-driven thermohaline circulation

Energy for the ocean circulation (and for the atmospheric circulation) comes from solar radiation and gravitational energy from the sun and moon. The amount of sunlight absorbed at the surface varies strongly with latitude, being greater at the equator than at the poles, and this engenders fluid motion in both the atmosphere and ocean that acts to redistribute heat from the equator towards the poles, thereby reducing the temperature gradients that would exist in the absence of fluid motion. Perhaps three quarters of this heat is carried in the atmosphere; the rest is carried in the ocean.

The atmosphere is heated from below, which leads to convection, the largest expression of which is the Hadley circulation. By contrast the ocean is heated from above, which tends to suppress convection. Instead ocean deep water is formed in polar regions where cold salty waters sink in fairly restricted areas. This is the beginning of the thermohaline circulation.

Oceanic currents are largely driven by the surface wind stress; hence the large-scale atmospheric circulation is important to understanding the ocean circulation. The Hadley circulation leads to Easterly winds in the tropics and Westerlies in mid-latitudes. This leads to slow equatorward flow throughout most of a subtropical ocean basin (the Sverdrup balance). The return flow occurs in an intense, narrow, poleward western boundary current. Like the atmosphere, the ocean is far wider than it is deep, and hence horizontal motion is in general much faster than vertical motion. In the southern hemisphere there is a continuous belt of ocean, and hence the mid-latitude westerlies force the strong Antarctic Circumpolar Current. In the northern hemisphere the land masses prevent this and the ocean circulation is broken into smaller gyres in the Atlantic and Pacific basins.

Coriolis Effect

The Coriolis effect results in a deflection of fluid flows (to the right in the Northern Hemisphere and left in the Southern Hemisphere). This has profound effects on the flow of the oceans. In particular it means the flow goes *around* high and low pressure systems, permitting them to persist for long periods of time. As a result, tiny variations in pressure can produce measurable currents. A slope of one part in one million in sea surface height, for example, will result in a current of 10 cm/s at mid-latitudes. The fact that the Coriolis effect is largest at the poles and weak at the equator results in sharp, relatively steady western boundary currents which are absent on eastern boundaries.

Ekman Transport

Ekman transport results in the net transport of surface water 90 degrees to the right of the wind in the Northern Hemisphere, and 90 degrees to the left of the wind in the Southern Hemisphere. As the wind blows across the surface of the ocean, it "grabs" onto a thin layer of the surface water. In turn, that thin sheet of water transfers motion energy to the thin layer of water under it, and so on. However, because of the Coriolis Effect, the direction of travel of the layers of water slowly move farther and farther to the right as they get deeper in the Northern Hemisphere, and to the left in the Southern Hemisphere. In most cases, the very bottom layer of water affected by the wind is at a depth of 100 m – 150 m and is traveling about 180 degrees, completely opposite of the direction that the wind is blowing. Overall, the net transport of water would be 90 degrees from the original direction of the wind.

Langmuir Circulation

Langmuir circulation results in the occurrence of thin, visible stripes, called windrows on the surface of the ocean parallel to the direction that the wind is blowing. If the wind is blowing with more than 3 m s^{-1}, it can create parallel windrows alternating upwelling and downwelling about 5–300 m apart. These windrows are created by adjacent ovular water cells (extending to about 6 m (20 ft) deep) alternating rotating clockwise and counterclockwise. In the convergence zones debris, foam and seaweed accumulates, while at the divergence zones plankton are caught and carried to the surface. If there are many plankton in the divergence zone fish are often attracted to feed on them.

Ocean–atmosphere Interface

Hurricane Isabel east of the Bahamas on 15 September 2003

At the ocean-atmosphere interface, the ocean and atmosphere exchange fluxes of heat, moisture and momentum.

Heat

The important heat terms at the surface are the sensible heat flux, the latent heat flux, the incoming solar radiation and the balance of long-wave (infrared) radiation. In general, the tropical oceans will tend to show a net gain of heat, and the polar oceans a net loss, the result of a net transfer of energy polewards in the oceans.

The oceans' large heat capacity moderates the climate of areas adjacent to the oceans, leading to a maritime climate at such locations. This can be a result of heat storage in summer and release in winter; or of transport of heat from warmer locations: a particularly notable example of this is Western Europe, which is heated at least in part by the north atlantic drift.

Momentum

Surface winds tend to be of order meters per second; ocean currents of order centimeters per second. Hence from the point of view of the atmosphere, the ocean can be considered effectively stationary; from the point of view of the ocean, the atmosphere imposes a significant wind stress on its surface, and this forces large-scale currents in the ocean.

Through the wind stress, the wind generates ocean surface waves; the longer waves have a phase velocity tending towards the wind speed. Momentum of the surface winds is transferred into the energy flux by the ocean surface waves. The increased roughness of the ocean surface, by the presence of the waves, changes the wind near the surface.

Moisture

The ocean can gain moisture from rainfall, or lose it through evaporation. Evaporative loss leaves the ocean saltier; the Mediterranean and Persian Gulf for example have strong evaporative loss; the resulting plume of dense salty water may be traced through the Straits of Gibraltar into the Atlantic Ocean. At one time, it was believed that evaporation/precipitation was a major driver of ocean currents; it is now known to be only a very minor factor.

Planetary Waves

Kelvin Waves

A Kelvin wave is any progressive wave that is channeled between two boundaries or opposing forces (usually between the Coriolis force and a coastline or the equator). There are two types, coastal and equatorial. Kelvin waves are gravity driven and non-dispersive. This means that Kelvin waves can retain their shape and direction over long periods of time. They are usually created by a sudden shift in the wind, such as the change of the trade winds at the beginning of the El Niño-Southern Oscillation.

Coastal Kelvin waves follow shorelines and will always propagate in a counterclockwise direction in the Northern hemisphere (with the shoreline to the right of the direction of travel) and clockwise in the Southern hemisphere.

Equatorial Kelvin waves propagate to the east in the Northern and Southern hemispheres, using the equator as a guide.

Kelvin waves are known to have very high speeds, typically around 2–3 meters per second. They have wavelengths of thousands of kilometers and amplitudes in the tens of meters.

Rossby Waves

Rossby waves, or planetary waves are huge, slow waves generated in the troposphere by temperature differences between the ocean and the continents. Their major restoring force is the change in Coriolis force with latitude. Their wave amplitudes are usually in the tens of meters and very large wavelengths. They are usually found at low or mid latitudes.

There are two types of Rossby waves, barotropic and baroclinic. Barotropic Rossby waves have the highest speeds and do not vary vertically. Baroclinic Rossby waves are much slower.

The special identifying feature of Rossby waves is that the phase velocity of each individual wave always has a westward component, but the group velocity can be in any direction. Usually the shorter Rossby waves have an eastward group velocity and the longer ones have a westward group velocity.

Climate Variability

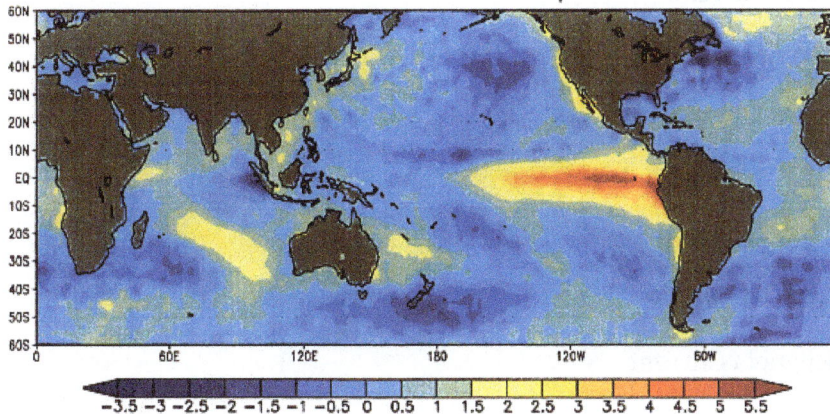

December 1997 chart of ocean surface temperature anomaly [°C] during the last strong El Niño

The interaction of ocean circulation, which serves as a type of heat pump, and biological effects such as the concentration of carbon dioxide can result in global climate changes on a time scale of decades. Known climate oscillations resulting from these interactions, include the Pacific decadal oscillation, North Atlantic oscillation, and Arctic oscillation. The oceanic process of thermohaline circulation is a significant component of heat redistribution across the globe, and changes in this circulation can have major impacts upon the climate.

La Niña–El Niño

Antarctic Circumpolar Wave

This is a coupled ocean/atmosphere wave that circles the Southern Ocean about every eight years. Since it is a wave-2 phenomenon (there are two peaks and two troughs in a latitude circle) at each fixed point in space a signal with a period of four years is seen. The wave moves eastward in the direction of the Antarctic Circumpolar Current.

Ocean Currents

Among the most important ocean currents are the:

- Antarctic Circumpolar Current

- Deep ocean (density-driven)

- Western boundary currents

  o Gulf Stream

  o Kuroshio Current

  o Labrador Current

  o Oyashio Current

  o Agulhas Current

  o Brazil Current

  o East Australia Current

- Eastern Boundary currents

  o California Current

  o Canary Current

  o Peru Current

  o Benguela Current

Antarctic Circumpolar

The ocean body surrounding the Antarctic is currently the only continuous body of water where there is a wide latitude band of open water. It interconnects the Atlantic, Pacific and Indian oceans, and provide an uninterrupted stretch for the prevailing westerly winds to significantly increase wave amplitudes. It is generally accepted that these prevailing winds are primarily responsible for the circumpolar current transport. This current is now thought to vary with time, possibly in an oscillatory manner.

Deep Ocean

In the Norwegian Sea evaporative cooling is predominant, and the sinking water mass, the North Atlantic Deep Water (NADW), fills the basin and spills southwards through crevasses in the submarine sills that connect Greenland, Iceland and Britain. It then flows along the western boundary of the Atlantic with some part of the flow moving eastward along the equator and then poleward into the ocean basins. The NADW is entrained into the Circumpolar Current, and can be traced into the Indian and Pacific basins. Flow from the Arctic Ocean Basin into the Pacific, however, is blocked by the narrow shallows of the Bering Strait.

Western Boundary

An idealised subtropical ocean basin forced by winds circling around a high pressure (anticyclonic) systems such as the Azores-Bermuda high develops a gyre circulation with slow steady flows towards the equator in the interior. As discussed by Henry Stommel, these flows are balanced in the region of the western boundary, where a thin fast polewards flow called a western boundary current develops. Flow in the real ocean is more complex, but the Gulf stream, Agulhas and Kuroshio are examples of such currents. They are narrow (approximately 100 km across) and fast (approximately 1.5 m/s).

Equatorwards western boundary currents occur in tropical and polar locations, e.g. the East Greenland and Labrador currents, in the Atlantic and the Oyashio. They are forced by winds circulation around low pressure (cyclonic).

Gulf Stream

The Gulf Stream, together with its northern extension, North Atlantic Current, is a powerful, warm, and swift Atlantic ocean current that originates in the Gulf of Mexico, exits through the Strait of Florida, and follows the eastern coastlines of the United States and Newfoundland to the northeast before crossing the Atlantic Ocean.

Kuroshio

The Kuroshio Current is an ocean current found in the western Pacific Ocean off the east coast of Taiwan and flowing northeastward past Japan, where it merges with the easterly drift of the North Pacific Current. It is analogous to the Gulf Stream in the Atlantic Ocean, transporting warm, tropical water northward towards the polar region.

Heat Flux

Heat Storage

Ocean heat flux is a turbulent and complex system which utilizes atmospheric measurement techniques such as eddy covariance to measure the rate of heat transfer expressed in the unit of joules or watts per second. Heat flux is the difference in temperature between two points through which the heat passes. Most of the Earth's heat storage is within its seas with smaller fractions of the heat transfer in processes such as evaporation, radiation, diffusion, or absorption into the sea floor. The majority of the ocean heat flux is through advection or the movement of the ocean's currents. For example, the majority of the warm water movement in the south Atlantic is thought to have originated in the Indian Ocean. Another example of advection is the nonequatorial Pacific heating which results from subsurface processes related to atmospheric anticlines. Recent warming observations of Antarctic Bottom Water in the Southern Ocean is of concern to ocean scientists because bottom water changes will effect currents, nutrients, and biota elsewhere. The international awareness of global warming has focused scientific research on this topic since the 1988 creation of the Intergovernmental Panel on Climate Change. Improved ocean observation, instrumentation, theory, and funding has increased scientific reporting on regional and global issues related to heat.

Sea Level Change

Tide gauges and satellite altimetry suggest an increase in sea level of 1.5–3 mm/yr over the past 100 years.

The IPCC predicts that by 2081-2100, global warming will lead to a sea level rise of 260 to 820 mm.

Rapid Variations

Tides

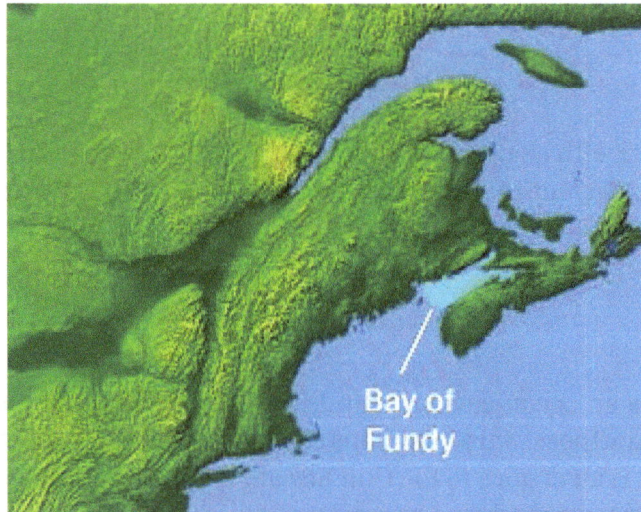

The Bay of Fundy is a bay located on the Atlantic coast of North America, on the northeast end of the Gulf of Maine between the provinces of New Brunswick and Nova Scotia.

The rise and fall of the oceans due to tidal effects is a key influence upon the coastal areas. Ocean tides on the planet Earth are created by the gravitational effects of the Sun and Moon. The tides produced by these two bodies are roughly comparable in magnitude, but the orbital motion of the Moon results in tidal patterns that vary over the course of a month.

The ebb and flow of the tides produce a cyclical current along the coast, and the strength of this current can be quite dramatic along narrow estuaries. Incoming tides can also produce a tidal bore along a river or narrow bay as the water flow against the current results in a wave on the surface.

Tide and Current (Wyban 1992) clearly illustrates the impact of these natural cycles on the lifestyle and livelihood of Native Hawaiians tending coastal fishponds. *Aia ke ola ka hana* meaning . . . *Life is in labor*.

Tidal resonance occurs in the Bay of Fundy since the time it takes for a large wave to travel from the mouth of the bay to the opposite end, then reflect and travel back to the mouth of the bay coincides with the tidal rhythm producing the world's highest tides.

As the surface tide oscillates over topography, such as submerged seamounts or ridges, it generates internal waves at the tidal frequency, which are known as internal tides.

Tsunamis

A series of surface waves can be generated due to large-scale displacement of the ocean water. These can be caused by sub-marine landslides, seafloor deformations due to earthquakes, or the impact of a large meteorite.

The waves can travel with a velocity of up to several hundred km/hour across the ocean surface, but in mid-ocean they are barely detectable with wavelengths spanning hundreds of kilometers.

Tsunamis, originally called tidal waves, were renamed because they are not related to the tides. They are regarded as shallow-water waves, or waves in water with a depth less than 1/20 their wavelength. Tsunamis have very large periods, high speeds, and great wave heights.

The primary impact of these waves is along the coastal shoreline, as large amounts of ocean water are cyclically propelled inland and then drawn out to sea. This can result in significant modifications to the coastline regions where the waves strike with sufficient energy.

The tsunami that occurred in Lituya Bay, Alaska on July 9, 1958 was 520 m (1,710 ft) high and is the biggest tsunami ever measured, almost 90 m (300 ft) taller than the Sears Tower in Chicago and about 110 m (360 ft) taller than the former World Trade Center in New York.

Surface Waves

The wind generates ocean surface waves, which have a large impact on offshore structures, ships, coastal erosion and sedimentation, as well as harbours. After their generation by the wind, ocean surface waves can travel (as swell) over long distances.

Wind Wave

In fluid dynamics, wind waves, or wind-generated waves, are surface waves that occur on the free surface of oceans, seas, lakes, rivers, and canals or even on small puddles and ponds. They result from the wind blowing over an area of fluid surface. Waves in the oceans can travel thousands of miles before reaching land. Wind waves range in size from small ripples, to waves over 100 ft (30 m) high.

When directly generated and affected by local winds, a wind wave system is called a wind sea. After the wind ceases to blow, wind waves are called swells. More generally, a swell consists of wind-generated waves that are not significantly affected by the local wind at that time. They have been generated elsewhere or some time ago. Wind waves in the ocean are called ocean surface waves.

Wind waves have a certain amount of randomness: subsequent waves differ in height, duration, and shape with limited predictability. They can be described as a stochastic process, in combination with the physics governing their generation, growth, propagation and decay—as well as governing the interdependence between flow quantities such as: the water surface movements, flow velocities and water pressure. The key statistics of wind waves (both seas and swells) in evolving sea states can be predicted with wind wave models.

Although waves are usually considered in the water seas of Earth, the hydrocarbon seas of Titan may also have wind-driven waves.

Wave Formation

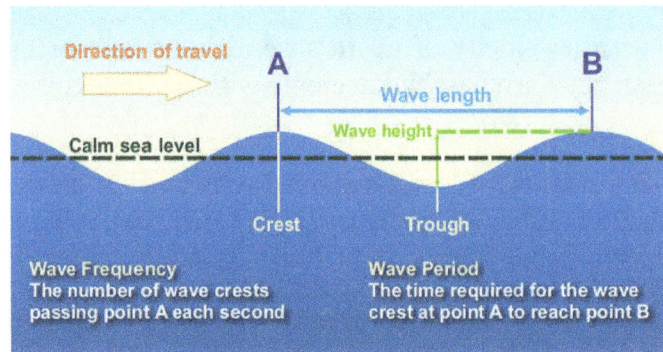

Aspects of a water wave

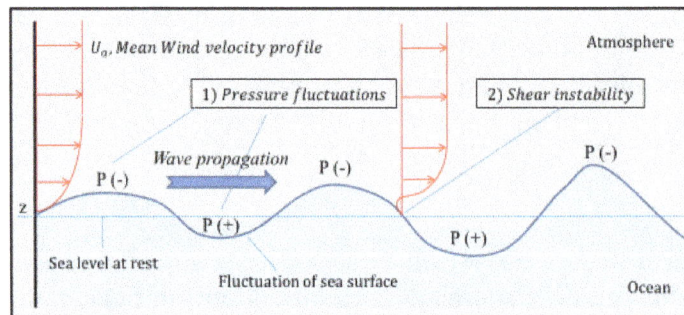

Wave formation

Water particle motion of a deep water wave

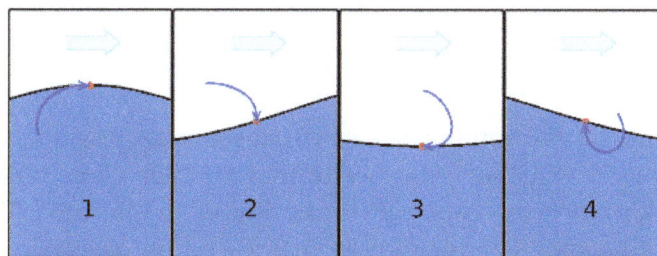

The phases of an ocean surface wave: 1. Wave Crest, where the water masses of the surface layer are moving horizontally in the same direction as the propagating wave front. 2. Falling wave. 3. Trough, where the water masses of the surface layer are moving horizontally in the opposite direction of the wave front direction. 4. Rising wave.

NOAA ship *Delaware II* in bad weather on Georges Bank

The great majority of large breakers seen on a beach result from distant winds. Five factors influence the formation of the flow structures in wind waves:

- Wind speed or strength relative to wave speed- the wind must be moving faster than the wave crest for energy transfer

- The uninterrupted distance of open water over which the wind blows without significant change in direction (called the *fetch*)

- Width of area affected by fetch

- Wind duration - the time over which the wind has blown over a given area

- Water depth

All of these factors work together to determine the size of wind waves and the structures of the flows within:

- Wave height (from high trough to crest)

- Wave length (from crest to crest)

- Wave period (time interval between arrival of consecutive crests at a stationary point)

- Wave propagation direction

A fully developed sea has the maximum wave size theoretically possible for a wind of a specific strength, duration, and fetch. Further exposure to that specific wind could only cause a loss of energy due to the breaking of wave tops and formation of "whitecaps". Waves in a given area typically have a range of heights. For weather reporting and for scientific analysis of wind wave statistics, their characteristic height over a period of time is usually expressed as *significant wave height*. This figure represents an average height of the highest one-third of the waves in a given time period (usually chosen somewhere in the range from 20 minutes to twelve hours), or in a specific wave or storm system. The significant wave height is also the value a "trained observer" (e.g. from a ship's crew) would estimate from visual observation of a sea state. Given the variability of wave height, the largest individual waves are likely to be somewhat less than twice the reported significant wave height for a particular day or storm.

• Sources of wind wave generation: Sea water wave is generated by many kinds of disturbances such as Seismic events, gravity, and crossing wind. The generation of wind wave is initiated by the disturbances of cross wind field on the surface of the sea water. Two major Mechanisms of surface wave formation by winds (a.k.a.'The Miles-Phillips Mechanism') and other sources (ex. earthquakes) of wave formation can explain the generation of wind waves.

However, if one set a flat water surface (Beaufort Point,0) and abrupt cross wind flows on the surface of the water, then the generation of surface wind waves can be explained by following two mechanisms which initiated by normal pressure fluctuations of turbulent winds and parallel wind shear flows.

• The mechanism of the surface wave generation by winds

1) Starts from "Fluctuations of wind" (O.M.Phillips) : the wind wave formation on water surface by wind is started by a random distribution of normal pressure acting on the water from the wind. By the mechanism developed by O.M. Phillips (in 1957), the water surface is initially at rest and wave generation is started by adding turbulent wind flows and then, by the fluctuations of the wind, normal pressure acting on the water surface. From this pressure fluctuation arise normal and tangential stresses to the surface water, which generates wave behavior on the water surface. It is assumed that:-

1. The water originally at rest.

2. The water is not viscid.

3. The water is irrotational.

4. There are random distribution of normal pressure to the water surface from the turbulent wind.

5. Correlations between air and water motions are neglected.

2) starts from "wind shear forces" on the water surface (J.W.Miles, applied to mainly 2D deep water gravity waves) ; John W. Miles suggested a surface wave generation mechanism which is initiated by turbulent wind shear flows Ua(y), based on the inviscid Orr-Sommerfeld equation in

1957. He found the energy transfer from wind to water surface as a wave speed, c is proportional to the curvature of the velocity profile of wind Ua"(y) at point where the mean wind speed is equal to the wave speed (Ua=c, where, Ua is the Mean turbulent wind speed). Since the wind profile Ua(y) is logarithmic to the water surface, the curvature Ua"(y) have negative sign at the point of Ua=c. This relations show the wind flow transferring its kinetic energy to the water surface at their interface, and arises wave speed, c.

the growth-rate can be determined by the curvature of the winds $((d^2 Ua)/(dz^2))$ at the steering height (Ua $(z=z\_h)=c$) for a given wind speed Ua {*Assumptions*; 1. 2D parallel shear flow, Ua(y) 2. incompressible, inviscid water / wind 3. irrotational water 4. slope of the displacement of surface is small}

Generally these wave formation mechanisms occur together on the ocean surface and arise wind waves and grows up to the fully developed waves.

For example,

If we suppose a very flat sea surface (Beaufort number, 0), and sudden wind flow blows steadily across the sea surface, physical wave generation process will be like;

1. Turbulent wind flows form random pressure fluctuations at the sea surface. Small waves with a few centimeters order of wavelengths is generated by the pressure fluctuations. (The Phillips mechanism)

2. The cross wind keep acting on the initially fluctuated sea surface, then the wave become larger. As the wave become larger, the pressure differences get larger along to the wave growing, then the wave growth rate is getting faster. Then the shear instability expedites the wave growing exponentially. (The Miles mechanism)

3. The interactions between the waves on the surface generate longer waves (Hasselmann et al., 1973) and the interaction will transfer wave energy from the shorter waves generated by the Miles mechanism to the waves have slightly lower frequencies than the frequency at the peak wave magnitudes, then finally the waves will be faster than the cross wind speed (Pierson & Moskowitz).

Conditions Necessary for a Fully Developed Sea at Given Wind Speeds, and the Parameters of the Resulting Waves					
Wind Conditions			Wave Size		
Wind Speed in One Direction	Fetch	Wind Duration	Average Height	Average Wavelength	Average Period and Speed
19 km/h (12 mph)	19 km (12 mi)	2 hr	0.27 m (0.89 ft)	8.5 m (28 ft)	3.0 sec 9.3 ft/sec
37 km/h (23 mph)	139 km (86 mi)	10 hr	1.5 m (4.9 ft)	33.8 m (111 ft)	5.7 sec 19.5 ft/sec
56 km/h (35 mph)	518 km (322 mi)	23 hr	4.1 m (13 ft)	76.5 m (251 ft)	8.6 sec 29.2 ft/sec
74 km/h (46 mph)	1,313 km (816 mi)	42 hr	8.5 m (28 ft)	136 m (446 ft)	11.4 sec 39.1 ft/sec
92 km/h (57 mph)	2,627 km (1,632 mi)	69 hr	14.8 m (49 ft)	212.2 m (696 ft)	14.3 sec 48.7 ft/sec

((NOTE: Most of the wave speeds calculated from the wave length divided by the period are proportional to sqrt (length). Thus, except for the shortest wave length, the waves follow the deep water theory described in the next section. The 28 ft long wave must be either in shallow water or between deep and shallow.))

Types of Wind Waves

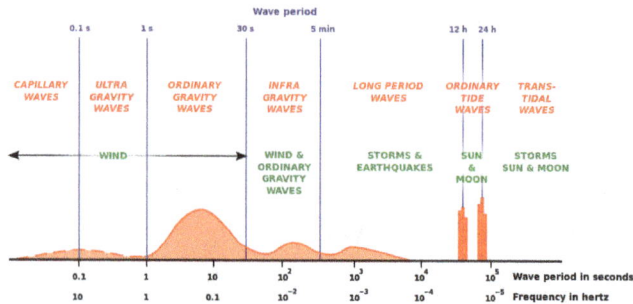

Classification of the spectrum of ocean waves according to wave period.

Surf on a rocky irregular bottom. Porto Covo, west coast of Portugal

Three different types of wind waves develop over time:

- Capillary waves, or ripples

- Seas

- Swells

Ripples appear on smooth water when the wind blows, but will die quickly if the wind stops. The restoring force that allows them to propagate is surface tension. Sea waves are larger-scale, often irregular motions that form under sustained winds. These waves tend to last much longer, even after the wind has died, and the restoring force that allows them to propagate is gravity. As waves propagate away from their area of origin, they naturally separate into groups of common direction and wavelength. The sets of waves formed in this way are known as swells.

Individual "rogue waves" (also called "freak waves", "monster waves", "killer waves", and "king waves") much higher than the other waves in the sea state can occur. In the case of the Draupner wave, its 25 m (82 ft) height was 2.2 times the significant wave height. Such waves are distinct from tides, caused by the Moon and Sun's gravitational pull, tsunamis that are caused by underwater earthquakes or landslides, and waves generated by underwater explosions or the fall of meteorites— all having far longer wavelengths than wind waves.

Yet, the largest ever recorded wind waves are common — not rogue — waves in extreme sea states. For example: 29.1 m (95 ft) high waves have been recorded on the RRS Discovery in a sea with 18.5 m (61 ft) significant wave height, so the highest wave is only 1.6 times the significant wave height. The biggest recorded by a buoy (as of 2011) was 32.3 m (106 ft) high during the 2007 typhoon Krosa near Taiwan.

Ocean waves can be classified based on: the disturbing force(s) that create(s) them; the extent to which the disturbing force(s) continue(s) to influence them after formation; the extent to which the restoring force(s) weaken(s) (or flatten) them; and their wavelength or period. Seismic Sea waves have a period of ~20 minutes, and speeds of 760 km/h (470 mph). Wind waves (deep-water waves) have a period of about 20 seconds.

Wave type	Typical wavelength	Disturbing force	Restoring force
Capillary wave	< 2 cm	Wind	Surface tension
Wind wave	60–150 m (200–490 ft)	Wind over ocean	Gravity
Seiche	Large, variable; a function of basin size	Change in atmospheric pressure, storm surge	Gravity
Seismic sea wave (tsunami)	200 km (120 mi)	Faulting of sea floor, volcanic eruption, landslide	Gravity
Tide	Half the circumference of Earth	Gravitational attraction, rotation of Earth	Gravity

The speed of all ocean waves is controlled by gravity, wavelength, and water depth. Most characteristics of ocean waves depend on the relationship between their wavelength and water depth. Wavelength determines the size of the orbits of water molecules within a wave, but water depth determines the shape of the orbits. The paths of water molecules in a wind wave are circular only when the wave is traveling in deep water. A wave cannot "feel" the bottom when it moves through water deeper than half its wavelength because too little wave energy is contained in the small circles below that depth. Waves moving through water deeper than half their wavelength are known as deep-water waves. On the other hand, the orbits of water molecules in waves moving through shallow water are flattened by the proximity of the sea surface bottom. Waves in water shallower than 1/20 their original wavelength are known as shallow-water waves. Transitional waves travel through water deeper than 1/20 their original wavelength but shallower than half their original wavelength.

In general, the longer the wavelength, the faster the wave energy will move through the water. For deep-water waves, this relationship is represented with the following formula:

$$C = L/T$$

where C is speed (celerity), L is wavelength, and T is time, or period (in seconds).

The speed of a deep-water wave may also be approximated by:

$$C = \sqrt{gL/2\pi}$$

where g is the acceleration due to gravity, 9.8 meters (32 feet) per second squared. Because g and π (3.14) are constants, the equation can be reduced to:

$$C = 1.251\sqrt{L}$$

when C is measured in meters per second and L in meters. Note that in both instances that wave speed is proportional to wavelength.

The speed of shallow-water waves is described by a different equation that may be written as:

$$C = \sqrt{gd} = 3.1\sqrt{d}$$

where C is speed (in meters per second), g is the acceleration due to gravity, and d is the depth of the water (in meters). The period of a wave remains unchanged regardless of the depth of water through which it is moving. As deep-water waves enter the shallows and feel the bottom, however, their speed is reduced and their crests "bunch up," so their wavelength shortens.

Wave Shoaling and Refraction

As waves travel from deep to shallow water, their shape alters (wave height increases, speed decreases, and length decreases as wave orbits become asymmetrical). This process is called shoaling.

Wave refraction is the process by which wave crests realign themselves as a result of decreasing water depths. Varying depths along a wave crest cause the crest to travel at different phase speeds, with those parts of the wave in deeper water moving faster than those in shallow water. This process continues until the crests become (nearly) parallel to the depth contours. Rays—lines normal to wave crests between which a fixed amount of energy flux is contained—converge on local shallows and shoals. Therefore, the wave energy between rays is concentrated as they converge, with a resulting increase in wave height.

Because these effects are related to a spatial variation in the phase speed, and because the phase speed also changes with the ambient current – due to the Doppler shift – the same effects of refraction and altering wave height also occur due to current variations. In the case of meeting an adverse current the wave *steepens*, i.e. its wave height increases while the wave length decreases, similar to the shoaling when the water depth decreases.

Wave Breaking

Big wave breaking

Breaking of a wave reaching the beach

Some waves undergo a phenomenon called "breaking". A breaking wave is one whose base can no longer support its top, causing it to collapse. A wave breaks when it runs into shallow water, or when two wave systems oppose and combine forces. When the slope, or steepness ratio, of a wave is too great, breaking is inevitable.

Individual waves in deep water break when the wave steepness—the ratio of the wave height H to the wavelength λ—exceeds about 0.07, so for $H > 0.07\,\lambda$. In shallow water, with the water depth small compared to the wavelength, the individual waves break when their wave height H is larger than 0.8 times the water depth h, that is $H > 0.8\,h$. Waves can also break if the wind grows strong enough to blow the crest off the base of the wave.

Three main types of breaking waves are identified by surfers or surf lifesavers. Their varying characteristics make them more or less suitable for surfing, and present different dangers.

- Spilling, or rolling: these are the safest waves on which to surf. They can be found in most areas with relatively flat shorelines. They are the most common type of shorebreak

- Plunging, or dumping: these break suddenly and can "dump" swimmers—pushing them to the bottom with great force. These are the preferred waves for experienced surfers. Strong offshore winds and long wave periods can cause dumpers. They are often found where there is a sudden rise in the sea floor, such as a reef or sandbar.

- Surging: these may never actually break as they approach the water's edge, as the water below them is very deep. They tend to form on steep shorelines. These waves can knock swimmers over and drag them back into deeper water.

Science of Waves

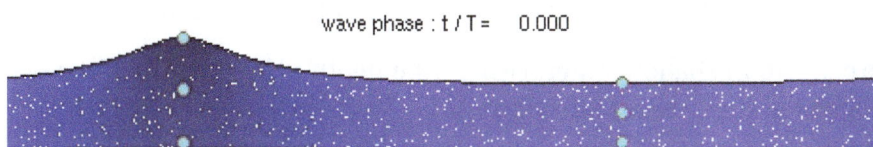

wave phase : t / T =   0.000

Stokes drift in shallow water waves (Animation)

Wind waves are mechanical waves that propagate. along the interface between water and air; the restoring force is provided by gravity, and so they are often referred to as surface gravity waves. As the wind blows, pressure and friction perturb the equilibrium of the water surface and transfer energy from the air to the water, forming waves. The initial formation of waves by the wind is described in the theory of Phillips from 1957, and the subsequent growth of the small waves has been modeled by Miles, also in 1957.

wave phase : t / T =   0.000

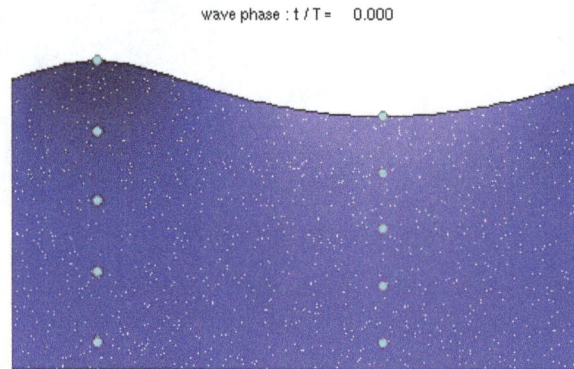

Stokes drift in a deeper water wave (Animation)

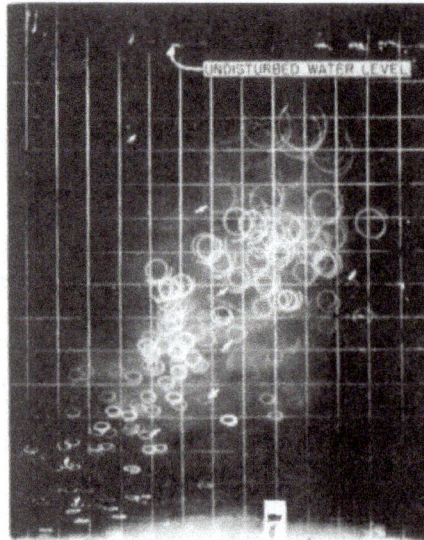

Photograph of the water particle orbits under a – progressive and periodic – surface gravity wave in a wave flume. The wave conditions are: mean water depth d = 2.50 ft (0.76 m), wave height H = 0.339 ft (0.103 m), wavelength λ = 6.42 ft (1.96 m), period T = 1.12 s.

In linear plane waves of one wavelength in deep water, parcels near the surface move not plainly up and down but in circular orbits: forward above and backward below (compared the wave propagation direction). As a result, the surface of the water forms not an exact sine wave, but more a trochoid with the sharper curves upwards—as modeled in trochoidal wave theory.

When waves propagate in shallow water, (where the depth is less than half the wavelength) the particle trajectories are compressed into ellipses.

In reality, for finite values of the wave amplitude (height), the particle paths do not form closed orbits; rather, after the passage of each crest, particles are displaced slightly from their previous positions, a phenomenon known as Stokes drift.

As the depth below the free surface increases, the radius of the circular motion decreases. At a depth equal to half the wavelength λ, the orbital movement has decayed to less than 5% of its value at the surface. The phase speed (also called the celerity) of a surface gravity wave is – for pure periodic wave motion of small-amplitude waves – well approximated by

$$c = \sqrt{\frac{g\lambda}{2\pi} \tanh\left(\frac{2\pi d}{\lambda}\right)}$$

where

c = phase speed;

λ = wavelength;

d = water depth;

g = acceleration due to gravity at the Earth's surface.

In deep water, where $d \geq \frac{1}{2}\lambda$, so $\frac{2\pi d}{\lambda} \geq \pi$ and the hyperbolic tangent approaches 1 , the speed c approximates

$$c_{\text{deep}} = \sqrt{\frac{g\lambda}{2\pi}}.$$

In SI units, with c_{deep} in m/s $c_{\text{deep}} \approx 1.25\sqrt{\lambda}$, when λ is measured in metres. This expression tells us that waves of different wavelengths travel at different speeds. The fastest waves in a storm are the ones with the longest wavelength. As a result, after a storm, the first waves to arrive on the coast are the long-wavelength swells. For intermediate and shallow water, the Boussinesq equations are applicable, combining frequency dispersion and nonlinear effects. And in very shallow water, the shallow water equations can be used.

If the wavelength is very long compared to the water depth, the phase speed (by taking the limit of c when the wavelength approaches infinity) can be approximated by

For intermediate and shallow water, the Boussinesq equations are applicable, combining frequency dispersion and nonlinear effects. And in very shallow water, the shallow water equations can be used.

If the wavelength is very long compared to the water depth, the phase speed (by taking the limit of c when the wavelength approaches infinity) can be approximated by

$$c_{\text{shallow}} = \lim_{\lambda \to \infty} c = \sqrt{gd}.$$

On the other hand, for very short wavelengths, surface tension plays an important role and the phase speed of these gravity-capillary waves can (in deep water) be approximated by

$$c_{\text{gravity-capillary}} = \sqrt{\frac{g\lambda}{2\pi} + \frac{2\pi S}{\rho\lambda}}$$

where

S = surface tension of the air-water interface;

ρ = density of the water.

When several wave trains are present, as is always the case in nature, the waves form groups. In deep water the groups travel at a group velocity which is half of the phase speed. Following a single wave in a group one can see the wave appearing at the back of the group, growing and finally disappearing at the front of the group.

As the water depth d decreases towards the coast, this will have an effect: wave height changes due to wave shoaling and refraction. As the wave height increases, the wave may become unstable when the crest of the wave moves faster than the trough. This causes *surf*, a breaking of the waves.

The movement of wind waves can be captured by wave energy devices. The energy density (per unit area) of regular sinusoidal waves depends on the water density ρ, gravity acceleration g and the wave height H (which, for regular waves, is equal to twice the amplitude, a):

$$E = \frac{1}{8}\rho g H^2 = \frac{1}{2}\rho g a^2.$$

The velocity of propagation of this energy is the group velocity.

Wind Wave Models

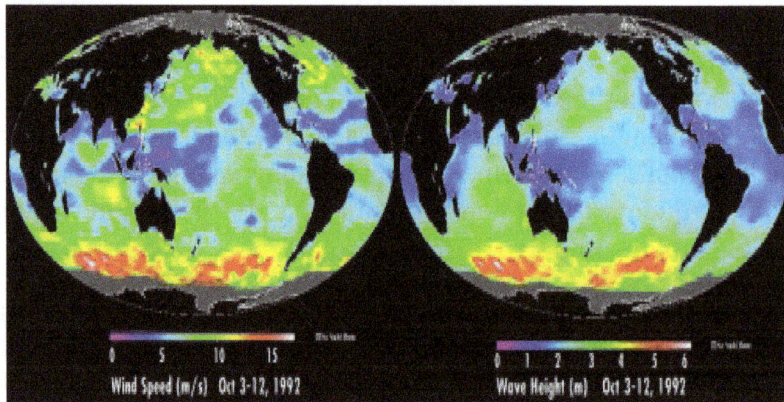

The image shows the global distribution of wind speed and wave height as observed by NASA's TOPEX/Poseidon's dual-frequency radar altimeter from October 3 to October 12, 1992. Simultaneous observations of wind speed and wave height are helping scientists to predict ocean waves. Wind speed is determined by the strength of the radar signal after it has bounced off the ocean surface and returned to the satellite. A calm sea serves as a good reflector and returns a strong signal; a rough sea tends to scatter the signals and returns a weak pulse. Wave height is determined by the shape of the return radar pulse. A calm sea with low waves returns a condensed pulse whereas a rough sea with high waves returns a stretched pulse. Comparing the two images above shows a high degree of correlation between wind speed and wave height. The strongest winds (33.6 mph; 54.1 km/h) and highest waves are found in the Southern Ocean. The weakest winds—shown as areas of magenta and dark blue—are generally found in the tropical Oceans.

Surfers are very interested in the wave forecasts. There are many websites that provide predictions of the surf quality for the upcoming days and weeks. Wind wave models are driven by more general weather models that predict the winds and pressures over the oceans, seas and lakes.

Wind wave models are also an important part of examining the impact of shore protection and beach nourishment proposals. For many beach areas there is only patchy information about the wave climate, therefore estimating the effect of wind waves is important for managing littoral environments.

Seismic Signals

Ocean water waves generate land seismic waves that propagate hundreds of kilometers into the land. These seismic signals usually have the period of 6 ± 2 seconds. Such recordings were first reported and understood in about 1900.

There are two types of seismic "ocean waves". The primary waves are generated in shallow waters by direct water wave-land interaction and have the same period as the water waves (10 to 16 seconds). The more powerful secondary waves are generated by the superposition of ocean waves of equal period traveling in opposite directions, thus generating standing gravity waves – with an associated pressure oscillation at half the period, which is not diminishing with depth. The theory for microseism generation by standing waves was provided by Michael Longuet-Higgins in 1950, after in 1941 Pierre Bernard suggested this relation with standing waves on the basis of observations.

Internal Waves

Internal waves can form at the boundary between water layers of different densities. These sub-surface waves are called internal waves. As is the case with ocean waves at the air-ocean interface, internal waves possess troughs, crests, wavelength, and period. Internal waves move very slowly because the density difference between the joined media is very small. Internal waves occur in the ocean at the base of the pycnocline, especially at the bottom edge of a steep thermocline. The wave height of internal waves may be greater than 30 meters (98 feet), causing the pycnocline to undulate slowly through a considerable depth. Their wavelength often exceeds 0.8 kilometres (0.50 mi) and their periods are typically 5 to 8 minutes. Internal waves are generated by wind energy, tidal energy, and ocean currents. Surface manifestations of internal waves have been photographed from space.

Internal waves may mix nutrients into surface water and trigger plankton blooms. They can also affect submarines and oil platforms.

Internal Tide

Internal tides are generated as the surface tides move stratified water up and down sloping topography, which produces a wave in the ocean interior. So internal tides are internal waves at a tidal frequency. The other major source of internal waves is the wind which produces internal waves near the inertial frequency. When a small water parcel is displaced from its equilibrium position, it will return either downwards due to gravity or upwards due to buoyancy. The water parcel will overshoot its original equilibrium position and this disturbance will set off an internal gravity wave. Munk (1981) notes, "Gravity waves in the ocean's interior are as common as waves at the sea surface-perhaps even more so, for no one has ever reported an interior calm."

Simple Explanation

The surface tide propagates as a wave, in which water parcels in the whole water column oscillate in the same direction at a given phase (i.e., in the trough or at the crest, Fig., top). At the simplest level, an internal wave can be thought of as an interfacial wave. If there are two levels in the ocean, such as a warm surface layer and cold deep layer separated by a thermocline,then motions on the interface are possible. The interface movement is large compared to surface movement. The restoring force for internal waves and tides is still gravity but its effect is reduced because the densities of the 2 layers are relatively similar compared to the large density difference at the air-sea interface. Thus larger displacements are possible inside the ocean than at the sea surface.

Simple interfacial internal wave

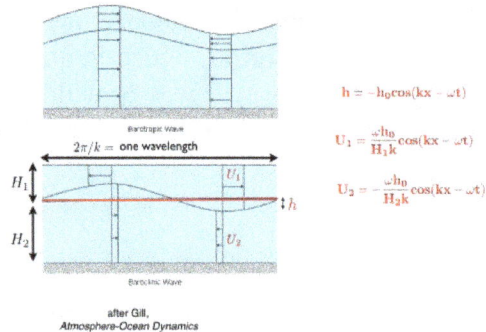

$$h = -h_0 \cos(kx - \omega t)$$

$$U_1 = \frac{\omega h_0}{H_1 k} \cos(kx - \omega t)$$

$$U_2 = -\frac{\omega h_0}{H_2 k} \cos(kx - \omega t)$$

after Gill,
Atmosphere-Ocean Dynamics

Water parcels in the whole water column move together with the surface tide (top), while shallow and deep waters move in opposite directions in an internal tide (bottom). The surface displacement and interface displacement are the same for a surface wave (top), while for an internal wave the surface displacements are very small, while the interface displacements are large (bottom). This figure is a modified version of one appearing in Gill (1982).

Tides occur mainly at diurnal and semidiurnal periods. The principal lunar semidiurnal constituent is known as M2 and generally has the largest amplitudes.

Location

The largest internal tides are generated at steep, midocean topography such as the Hawaiian Ridge, Tahiti, the Macquarie Ridge, and submarine ridges in the Luzon Strait. Continental slopes such as the Australian North West Shelf also generate large internal tides. These internal tide may propagate onshore and dissipate much like surface waves. Or internal tides may propagate away from the topography into the open ocean. For tall, steep, midocean topography, such as the Hawaiian Ridge, it is estimated that about 85% of the energy in the internal tide propagates away into the deep ocean with about 15% of its energy being lost within about 50 km of the generation site. The lost energy contributes to turbulence and mixing near the generation sites. It is not clear where the energy that leaves the generation site is dissipated, but there are 3 possible processes: 1) the internal tides scatter and/or break at distant midocean topography, 2) interactions with other internal waves remove energy from the internal tide, or 3) the internal tides shoal and break on continental shelves.

Propagation and Dissipation

Briscoe (1975)noted that "We cannot yet answer satisfactorily the questions: 'where does the internal wave energy come from, where does it go, and what happens to it along the way?'" Although technological advances in instrumentation and modeling have produced greater knowledge of internal tide and near-inertial wave generation, Garrett and Kunze (2007) observed 33 years later that "The fate of the radiated [large-scale internal tides] is still uncertain. They may scatter into [smaller scale waves] on further encounter with islands or the rough seafloor , or transfer their energy to smaller-scale internal waves in the ocean interior " or "break on distant continental slopes ". It is now known that most of the internal tide energy generated at tall, steep midocean topography radiates away as large-scale internal waves. This radiated internal tide energy is one of the main sources of energy into the deep ocean, roughly half of the wind energy input . Broader interest in internal tides is spurred by their impact on the magnitude and spatial inhomogeneity of mixing, which in turn has first order effect on the meridional overturning circulation.

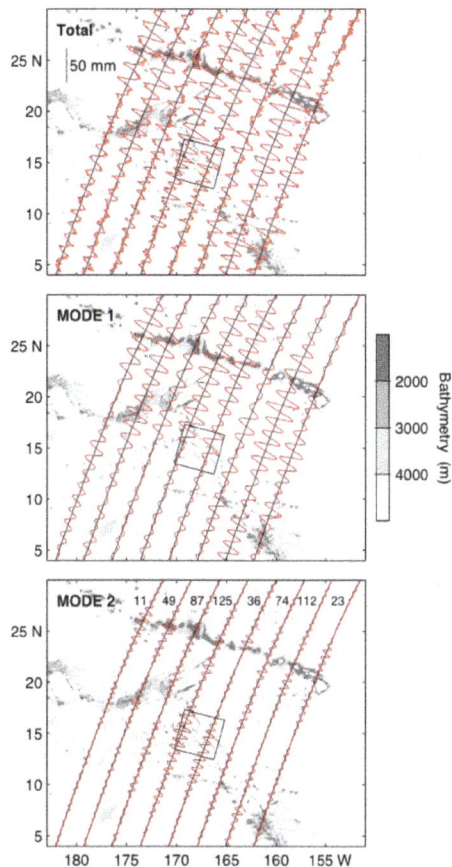

The internal tide sea surface elevation that is in phase with the surface tide (i.e., crests occur in a certain spot at a certain time that are both the same relative to the surface tide) can be detected by satellite (top). (The satellite track is repeated about every 10 days and so M2 tidal signals are shifted to longer periods due to aliasing.) The longest internal tide wavelengths are about 150 km near Hawaii and the next longest waves are about 75 km long. The surface displacements due to the internal tide are plotted as wiggly red lines with amplitudes plotted perpendicular to the satellite groundtracks (black lines). Figure is adapted from Johnston et al. (2003).

The internal tidal energy in one tidal period going through an area perpendicular to the direction of propagation is called the energy flux and is measured in Watts/m. The energy flux at one point can be summed over depth- this is the depth-integrated energy flux and is measured in Watts/m. The Hawaiian Ridge produces depth-integrated energy fluxes as large as 10 kW/m. The longest wavelength waves are the fastest and thus carry most of the energy flux. Near Hawaii, the typical wavelength of the longest internal tide is about 150 km while the next longest is about 75 km. These waves are called mode 1 and mode 2, respectively. Although Fig. shows there is no sea surface expression of the internal tide, there actually is a displacement of a few centimeters. These sea surface expressions of the internal tide at different wavelengths can be detected with the Topex/Poseidon or Jason-1 satellites. Near 15 N, 175 W on the Line Islands Ridge, the mode-1 internal tides scatter off the topography, possibly creating turbulence and mixing, and producing smaller wavelength mode 2 internal tides.

The inescapable conclusion is that energy is lost from the surface tide to the internal tide at midocean topography and continental shelves, but the energy in the internal tide is not necessarily lost in the same place. Internal tides may propagate thousands of kilometers or more before breaking and mixing the abyssal ocean.

Abyssal Mixing and Meridional Overturning Circulation

The importance of internal tides and internal waves in general relates to their breaking, energy dissipation, and mixing of the deep ocean. If there were no mixing in the ocean, the deep ocean would be a cold stagnant pool with a thin warm surface layer. While the meridional overturning circulation (also referred to as the thermohaline circulation) redistributes about 2 PW of heat from the tropics to polar regions, the energy source for this flow is the interior mixing which is comparatively much smaller- about 2 TW. Sandstrom (1908) showed a fluid which is both heated and cooled at its surface cannot develop a deep overturning circulation. Most global models have incorporated uniform mixing throughout the ocean because they do not include or resolve internal tidal flows.

However, models are now beginning to include spatially variable mixing related to internal tides and the rough topography where they are generated and distant topography where they may break. Wunsch and Ferrari (2004) describe the global impact of spatially inhomogeneous mixing near midocean topography: "A number of lines of evidence, none complete, suggest that the oceanic general circulation, far from being a heat engine, is almost wholly governed by the forcing of the wind field and secondarily by deep water tides... The now inescapable conclusion that over most of the ocean significant 'vertical' mixing is confined to topographically complex boundary areas implies a potentially radically different interior circulation than is possible with uniform mixing. Whether ocean circulation models... neither explicitly accounting for the energy input into the system nor providing for spatial variability in the mixing, have any physical relevance under changed climate conditions is at issue." There is a limited understanding of "the sources controlling the internal wave energy in the ocean and the rate at which it is dissipated" and are only now developing some "parameterizations of the mixing generated by the interaction of internal waves, mesoscale eddies, high-frequency barotropic fluctuations, and other motions over sloping topography."

Internal Tides at the Beach

Internal tides may also dissipate on continental slopes and shelves or even reach within 100 m of the beach. Internal tides bring pulses of cold water shoreward and produce large vertical temperature differences. When surface waves break, the cold water is mixed upwards, making the water cold for surfers, swimmers, and other beachgoers. Surface waters in the surf zone can change by about 10 °C in about an hour.

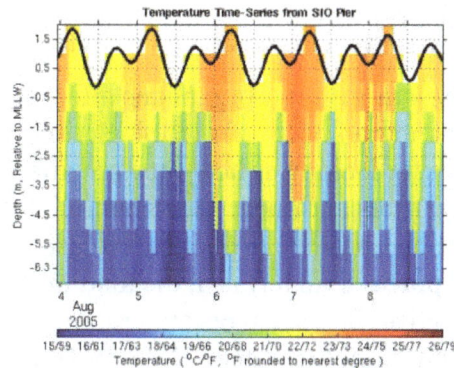

The internal tide produces large vertical differences in temperature at the research pier at the Scripps Institution of Oceanography. The black line shows the surface tide elevation relative to mean lower low water (MLLW). Figure provided by Eric Terrill, Scripps Institution of Oceanography with funding from the U.S. Office of Naval Research

Internal Tides, Internal Mixing, and Biological Enhancement

Internal tides generated by tidal semidiurnal currents impinging on steep submarine ridges in island passages, ex: Mona Passage, or near the shelf edge, can enhance turbulent dissipation and internal mixing near the generation site. The development of Kelvin-Helmholtz instability during the breaking of the internal tide can explain the formation of high diffusivity patches that generate a vertical flux of nitrate (NO_3^-) into the photic zone and can sustain new production locally. Another mechanism for higher nitrate flux at spring tides results from pulses of strong turbulent dissipation associated with high frequency internal soliton packets. Some internal soliton packets are the result of the nonlinear evolution of the internal tide.

Chemical Oceanography

Chemical oceanography is the study of ocean chemistry: the behavior of the chemical elements within the Earth's oceans. The ocean is unique in that it contains - in greater or lesser quantities - nearly every element in the periodic table.

Much of chemical oceanography describes the cycling of these elements both within the ocean and with the other spheres of the Earth system. These cycles are usually characterised as quantitative fluxes between c onstituent reservoirs defined within the ocean system and as residence times within the ocean. Of particular global and climatic significance are the cycles of the biologically active elements such as carbon, nitrogen, and phosphorus as well as those of some important trace elements such as iron.

Another important area of study in chemical oceanography is the behaviour of isotopes and how they can be used as tracers of past and present oceanographic and climatic processes. For example, the incidence of ^{18}O (the heavy isotope of oxygen) can be used as an indicator of polar ice sheet extent, and boron isotopes are key indicators of the pH and CO_2 content of oceans in the geologic past.

Biological Oceanography

Biological oceanography is the study of how organisms affect and are affected by the physics, chemistry, and geology of the oceanographic system. Biological oceanography mostly focuses on the microorganisms within the ocean; looking at how they are affected by their environment and how that affects larger marine creatures and their ecosystem. Biological oceanography is similar to marine biology, but is different because of the perspective used to study the ocean. Biological oceanography takes a bottom up approach (in terms of the food web), while marine biology studies the ocean from a top down perspective. Biological oceanography mainly focuses on the ecosystem of the ocean with an emphasis on plankton: their diversity (morphology, nutritional sources, motility, and metabolism); their productivity and how that plays a role in the global carbon cycle; and their distribution (predation and life cycle). Biological oceanography also investigates the role of microbes in food webs, and how humans impact the ecosystems in the oceans.

History

H.M.S. CHALLENGER UNDER SAIL, 1874.

HMS *Challenger* during its pioneer expedition of 1872–76

The Challenger Expedition was pivotal to biological oceanography and oceanography in general. The Challenger Expedition was headed by Charles Wyville Thomson in 1872-1876 The expedition also included two other naturalists, Henry N. Moseley and John Murray. Prior to the expedition, the ocean was, although interesting to many, considered an unpredictable and mostly life-less body of water and this expedition made them rethink this stance on the ocean This expedition was at the behest of The Royal Society in order to see if they would be able to lay cables at the bottom of the ocean. They also brought the equipment to collect data about the biological, chemical, and geological properties of the ocean in a systematic way. They mapped the oceanic sediment and collected data The data collected in this voyage proved that there was life in deep waters (5500 meters) and that the composition of water in the ocean is consistent.

Acoustical oceanography

Acoustical oceanography is the use of underwater sound to study the sea, its boundaries and its contents. Physical oceanographers' studies acoustical oceanography which includes topics on underwater acoustics, sound transmissions, etc.

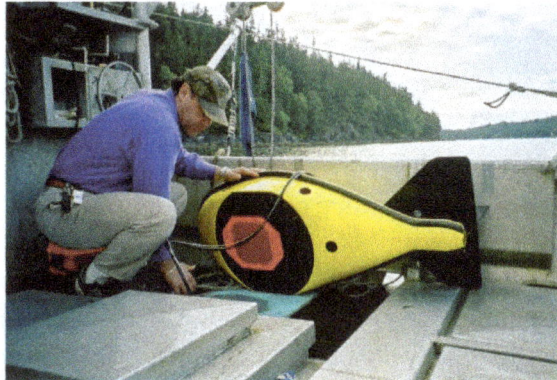

A 38 kHz hydroacoustic tow fin used to conduct acoustic surveys by NOAA. Alaska, Southeast.

History

Important contributions to acoustical oceanography have been made by:

- Leonid Brekhovskikh
- Walter Munk
- Hank Medwin
- John L Spiesberger
- C C Leroy
- David E. Weston
- D. Van Holliday
- Charles Greenlaw

Equipment Used

The earliest and most widespread use of sound and sonar technology to study the properties of the sea is the use of an rainbow echo sounder to measure water depth. Sounders were the devices used that mapped the many miles of the Santa Barbara Harbor ocean floor until 1993.

Fathometers measure the depth of the waters. It works by electronically sending sounds from ships, therefore also receiving the sound waves that bounces back from the bottom of the ocean. A paper chart moves through the fathometer and is calibrated to record the depth.

As technology advances, the development of high resolution sonars in the second half of the 20th century made it possible to not just detect underwater objects but to classify them and even image them. Electronic sensors are now attached to ROVs since nowadays, ships or robot submarines have Remotely Operated Vehicles (ROVs). There are cameras attached to these devices giving out accurate images. The oceanographers are able to get a clear and precise quality of pictures. The 'pictures' can also be sent from sonars by having sound reflected off ocean surroundings. Often-times sound waves reflect off animals, giving information which can be documented into deeper animal behaviour studies.

Paleoceanography

Paleoceanography is the study of the history of the oceans in the geologic past with regard to circulation, chemistry, biology, geology and patterns of sedimentation and biological productivity. Paleoceanographic studies using environment models and different proxies enable the scientific community to assess the role of the oceanic processes in the global climate by the re-construction of past climate at various intervals. Paleoceanographic research is also intimately tied to paleoclimatology.

Source and Methods of Information

Paleoceanography makes use of so-called proxy methods as a way to infer information about the past state and evolution of the worlds oceans. Several geochemical proxy tools include long-chain organic molecules (e.g. alkenones), stable and radioactive isotopes, and trace metals [Henderson, 2002]. Additionally, sediment cores can also be useful; the field of paleoceanography is closely related to sedimentology and paleontology.

Sea-surface Temperature

Sea-surface temperature (SST) records can be extracted from deep-sea sediment cores using oxygen isotope ratios and the ratio of magnesium to calcium (Mg/Ca) in shell secretions from plankton, from long-chain organic molecules such as alkenone, from tropical corals near the sea surface, and from mollusk shells .

Oxygen isotope ratios ($\delta^{18}O$) are useful in reconstructing SST because of the influence temperature has on the isotope ratio. Plankton take up oxygen in building their shells and will be less enriched in their $\delta^{18}O$ when formed in warmer waters, provided they are in thermodynamic equilibrium with the seawater [Urey, 1947]. When these shells precipitate, they sink and form sediments on the ocean floor whose $\delta^{18}O$ can be used to infer past SSTs [Emiliani, 1955]. Oxygen isotope ratios are not perfect proxies, however. The volume of ice trapped in continental ice sheets can have an impact of the $\delta^{18}O$. Freshwater characterized by lower values of $\delta^{18}O$ becomes trapped in the continental ice sheets, so that during glacial periods seawater $\delta^{18}O$ is elevated and calcite shells formed during these times will have a larger $\delta^{18}O$ value [Olausson, 1965; Shackleton, 1967].

The substitution of magnesium in place of calcium in $CaCO_3$ shells can be used as a proxy for the SST in which the shells formed. Mg/Ca ratios have several other influencing factors other than temperature, such as vital effects, shell-cleaning, and postmortem and post-depositional dissolution effects, to name a few . Other influences aside, Mg/Ca ratios have successfully quantified the tropical cooling that occurred during the last glacial period [Lea et al., 2003].

Alkenone is a long-chain, complex organic molecule produced by photosynthetic algae which is temperature sensitive and can be extracted from marine sediments. Use of alkenone represents a more direct relationship between SST and algae and does not rely on knowing biotic and physical-chemical thermodynamic relationships needed in $CaCO_3$ studies [Herbert, 2003]. Another advantage of using alkenone is that it is a product of photosynthesis and necessitates formation in the sunlight of the upper surface layers. As such, it is a better record of near-surface SST [Cornin].

Bottom-water Temperature

The most commonly used proxy to infer deep-sea temperature history are the Mg/Ca ratios in benthic foraminifera and ostracodes. The temperatures inferred from the Mg/Ca ratios have confirmed an up to 3 °C cooling of the deep ocean during the late Pleistocene glacial periods .

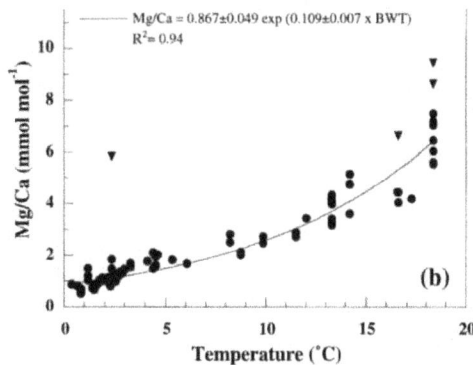

From Lear et al. Figure 8-B: Cibicidoides spp. Mg/Ca-temperature calibration. Compilation of corrected FAAS, ICP-AES, and ICP-MS data for C. wuellerstorfi, C. pachyderma, C. compressus, and a wuellerstorfi-like Cibicidoides. Triangles represent rejected data lying in the upper 5 percentile of the standard error.

One notable study is that by Lear et al. who worked to calibrate bottom water temperature to Mg/Ca ratios in 9 locations covering a variety of depths from up to six different benthic foraminifera (depending on location). The authors found an equation calibrating bottom water temperature ot Mg/Ca ratios that takes on an exponential form:

$$Mg/Ca = 0.867 +/- 0.049*exp\{0.109 +/- 0.007*BWT\}$$

where Mg/Ca is the Mg/Ca ratio found in the benthic foraminifera and BWT is the bottom water temperature.

Salinity

Salinity is a more challenging quantity to infer from paleorecords. Deuterium excess in core records can provide a better inference of sea-surface salinity than oxygen isotopes, and certain species, such as diatoms, can provide a semiquantitative salinity record due to the relative abundances of diatoms that are limited to certain salinity regimes [Bauch and Polyakova, 2003].

Ocean Circulation

Several proxy methods have been used to infer past ocean circulation and changes to it. They include carbon isotope ratios, cadmium/calcium (Cd/Ca) ratios, protactinium/thorium isotopes ([231]Pa and [230]Th), radiocarbon activity (δ^{14}C), neodymium isotopes ([143]Nd and [144]Nd), and sortable silt (fraction of deep-sea sediment between 10 and 63 µm) . Carbon isotope and cadmium/calcium ratio proxies are used because variability in their ratios is due partly to changes in bottom-water chemistry, which is in turn related the source of deep-water formation [e.g. Oppo and Lehman 1993; Lehman and Keigwin 1992]. These ratios, however, are influenced by biological, ecological, and geochemical processes which complicate circulation inferences.

All proxies included are useful in inferring the behavior of the meridional overturning circulation . For example, McManus et al. used protactinium/thorium isotopes (^{231}Pa and ^{230}Th) to show that the Atlantic Meridional Overturning Circulation had been nearly (or completely) shut off during the last glacial period. ^{231}Pa and ^{230}Th are both formed from the radioactive decay of dissolved uranium in seawater, with ^{231}Pa able to remain supported in the water column longer than ^{230}Th: ^{231}Pa has a residence time ~100–200 years while ^{230}Th has one ~20–40 years [McManus et al., 2004]. In today's Atlantic Ocean and current overturning circulation, ^{230}Th transport to the Southern Ocean is minimal due to its short residence time, and ^{231}Pa transport is high. This results in relatively low ^{231}Pa / ^{230}Th ratios found by McManus et al. in a core at 33N 57W, and a depth of 4.5 km. When the overturning circulation shuts down (as hypothesized) during glacial periods, the ^{231}Pa / ^{230}Th ratio becomes elevated due to the lack of removal of ^{231}Pa to the Southern Ocean. McManus et al. also note a small raise in the ^{231}Pa / ^{230}Th ratio during the Younger Dryas event, another period in climate history thought to have experienced a weakening overturning circulation.

Acidity, pH, and Alkalinity

Boron isotope ratios (δ^{11}B) can be used to infer both recent as well as millennial time scale changes in the acidity, pH, and alkalinity of the ocean, which is mainly forced by atmospheric CO_2 concentrations and bicarbonate ion concentration in the ocean. δ^{11}B has been identified in corals from the southwestern Pacific to vary with ocean pH, and shows that climate variabilities such as the Pacific decadal oscillation (PDO) can modulate the impact of ocean acidification due to rising atmospheric CO_2 concentrations .

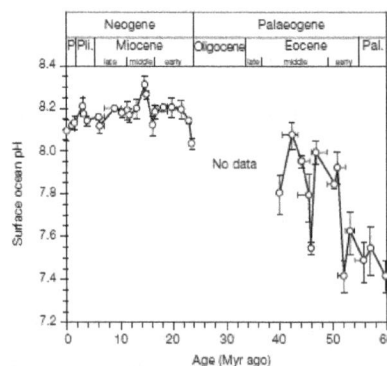

Figure 1 Sea surface pH for the past 60 Myr. Vertical error bars result from analytical error in determining δ^{11}B$_{cc}$. Horizontal error bars represent the higher and lower biostratigraphical datums that constrain each sample (Table 1). Timescale according to ref. 49. P, Pleistocene; Pli, Pliocene; Pal., Palaeocene.

Screen Shot of Figure 1 from Pearson and Palmer showing surface ocean pH as reconstructed from Boron isotope analysis for the past 60 million years.

Another application of δ^{11}B in plankton shells can be used as an indirect proxy for atmospheric CO_2 concentrations over the past several million years [Pearson and Palmer, 2000].

References

- Hamblin, W. Kenneth; Christiansen, Eric H. (1998). Earth's Dynamic Systems (8th ed.). Upper Saddle River: Prentice-Hall. ISBN 0-13-018371-7.

- Rousmaniere, John (1989). The Annapolis Book of Seamanship (2nd revised ed.). Simon & Schuster. ISBN 0-671-67447-1.

- Johnston, T. M. S.; M. A. Merrifield (2003). "Internal tide scattering at seamounts, ridges and islands". J. Geophys. Res. 108. (C6) 3126 (C6): 3180. Bibcode:2003JGRC..108.3180J. doi:10.1029/2002JC001528.

- St. Laurent; L. C.; C. Garrett (2002). "The Role of Internal Tides in Mixing the Deep Ocean". J. Phys. Oceanogr. 32 (10): 2882–2899. Bibcode:2002JPO....32.2882S. doi:10.1175/1520-0485(2002)032<2882:TROITI>2.0. CO;2. ISSN 1520-0485.

- MacKinnon, J. A.; K. B. Winters (2005). "Subtropical catastrophe: Significant loss of low-mode tidal energy at 28.9 degrees". Geophys. Res. Lett. 32. L15605 (15): L15605. Bibcode:2005GeoRL..3215605M. doi:10.1029/2005GL023376.

- Nash, J. D., E. Kunze, J.M. Toole, and R.W. Schmitt (2004). "Internal tide reflection and turbulent mixing on the continental slope". J. Phys. Oceanogr. 34 (5): 1117–1134. Bibcode:2004JPO....34.1117N. doi:10.1175/1520-0485(2004)034<1117:ITRATM>2.0.CO;2. ISSN 1520-0485.

- Wunsch, C.; R. Ferrari (2004). "Vertical mixing, energy, and the general circulation of the ocean". Annu. Rev. Fluid Mech. 36 (1): 281–314. Bibcode:2004AnRFM..36..281W. doi:10.1146/annurev.fluid.36.050802.122121.

- Alfonso-Sosa, E. (2002). Variabilidad temporal de la producción primaria fitoplanctonica en la estación CaTS (Caribbean Time-Series Station): Con énfasis en el impacto de la marea interna semidiurna sobre la producción. (PDF). Ph. D. Dissertation. Department of Marine Sciences, University of Puerto Rico, Mayagüez, Puerto Rico. UMI publication AAT 3042382. p. 407. Retrieved 2014-08-25.

- Alfonso-Sosa, E., J. M. Lopez, J. E. Capella, A. Dieppa and J. Morell (2002). "Internal Tide-induced Variations in Primary Productivity and Optical Properties in the Mona Passage, Puerto Rico" (PDF). Retrieved 2015-01-01.

- Sharples, J., J. F. Tweddle, J. A. M. Green, M. R. Palmer, Y. Kim, A. E. Hickman,P. M. Holligan, C. M. Moore, T. P. Rippeth, J. H. Simpson and V. Krivtsov (2007). "Spring–neap modulation of internal tide mixing and vertical nitrate fluxes at a shelf edge in summer" (PDF). Limnol. Oceanogr. 52 (5): 1735–1747. doi:10.4319/lo.2007.52.5.1735. Retrieved 2014-08-25.

- Lear, C.H., Y. Rosenthal, N. Slowey. 2002. Benthic foraminiferal Mg/Ca-paleothermometry: A revised core-top calibration. Geochimica et Cosmochimica Acta. Vol. 66. pp. 3375–3387.

Tools and Methods of Oceanography

Measuring large portions of the ocean for a better understanding of the temperature of the ocean is known as ocean acoustic tomography while the study of the depth of the lake and ocean floor is bathymetry. The chapter serves as a source to understand the major categories related to oceanography, providing the reader with a detailed insight on the methods used in oceanography.

Argo (Oceanography)

Argo is a system for observing temperature, salinity, and currents in the Earth's oceans which has been operational since the early 2000s. The real-time data it provides is used in climate and oceanographic research. A special research interest is to quantify the ocean heat content (OHC).

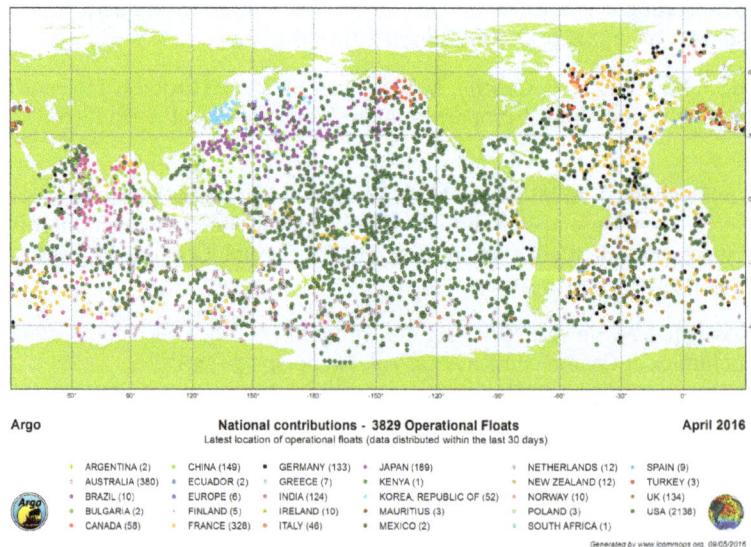

Argo **National contributions - 3829 Operational Floats** **April 2016**
Latest location of operational floats (data distributed within the last 30 days)

ARGENTINA (2)	CHINA (149)	GERMANY (133)	JAPAN (189)	NETHERLANDS (12)	SPAIN (9)
AUSTRALIA (380)	ECUADOR (2)	GREECE (7)	KENYA (1)	NEW ZEALAND (12)	TURKEY (3)
BRAZIL (10)	EUROPE (6)	INDIA (124)	KOREA, REPUBLIC OF (52)	NORWAY (10)	UK (134)
BULGARIA (2)	FINLAND (5)	IRELAND (10)	MAURITIUS (3)	POLAND (3)	USA (2138)
CANADA (58)	FRANCE (328)	ITALY (46)	MEXICO (2)	SOUTH AFRICA (1)	

Generated by www.jcommops.org, 09/05/2016

The distribution of active floats in the Argo array, colour coded by country that owns the float, as of the end of April 2016.

Argo consists of a fleet of almost 4000 drifting profiling floats deployed worldwide. Each Argo float weighs 20–30 kg. Profiling floats are commonly used in oceanography and become "Argo floats" only when they are deployed in conformity with the Argo data policy. In most cases probes drift at a depth of 1000 metres (the so-called parking depth) and, every 10 days, by changing their buoyancy, dive to a depth of 2000 metres and then move to the sea-surface, measuring conductivity and temperature profiles as well as pressure. From these, salinity and density can be calculated. Seawater density is important in determining large-scale motions in the ocean.

Average current velocities at 1000 metres are directly measured by the distance and direction a float drifts while parked at that depth, which is determined by GPS or Argos system positions at the surface. The data are transmitted to shore via satellite, and are freely available to everyone, without restrictions.

The Argo program is named after the Greek mythical ship *Argo* to emphasize the complementary relationship of Argo with the Jason satellite altimeters.

International Collaboration

The Argo program is a collaborative partnership of more than 30 nations from all continents (most shown on the graphic map in this article) to provide a seamless global array allowing any country to explore the ocean environment. Argo is a component of the Global Ocean Observing System (GOOS). Argo is coordinated by the Argo Steering Team – an international body of scientists and technical experts that meets once per year. The Argo data stream is managed by the Argo Data Management Team. Overall coordination is provided through the Argo Information Centre, an office belonging to the Intergovernmental Oceanographic Commission which also coordinates GOOS, and the World Meteorological Organization. Argo is also supported by GEO (the Group on Earth Observations), and has been endorsed since its early beginnings by the World Climate Research Programme's CLIVAR Project (Variability and predictability of the ocean-atmosphere system), and by the Global Ocean Data Assimilation Experiment (GODAE OceanView).

An animation for children was created recently by IMOS (Integrated Marine Observing Strategy, Australia) showing how Argo works.

History

A program called Argo was first proposed at OceanObs 1999 which was a conference organised by international agencies with the aim of creating a coordinated approach to ocean observations. The original Argo prospectus was created by a small group of scientists, chaired by Dean Roemmich, who described a program that would have a global array of about 3000 floats in place by sometime in 2007. The 3000-float array was achieved in November 2007 and was global. The Argo Steering Team met for the first time in 1999 in Maryland (USA) and outlined the principles of global data sharing. The Argo Steering Team made a 10-year report to OceanObs-2009 and received suggestions on how the array might be improved. These suggestions included enhancing the array at high latitudes, in marginal seas (such as the Gulf of Mexico and the Mediterranean) and along the equator, improved observation of strong boundary currents (such as the Gulf Stream and Kuroshio), extension of observations into deep water and the addition of sensors for monitoring biological and chemical changes in the oceans. In November 2012 an Indian float in the Argo array gathered the one-millionth profile (twice the number collected by research vessels during all of the 20th century) an event that was reported in several press releases. In 2014 the Bio-Argo program was expanding rapidly.

Float Design and Operation

The critical capability of an Argo float is its ability to rise and descend in the ocean on a programmed schedule. The floats do this by changing their effective density. The density of any object is given by its mass divided by its volume. The Argo float keeps its mass constant, but by altering its volume, it changes its density. To do this, mineral oil is forced out of the float's pressure case and expands a rubber bladder at the bottom end of the float. As the bladder expands, the float becomes less dense than seawater and rises to the surface. Upon finishing its tasks at the surface, the float withdraws the oil and descends again.

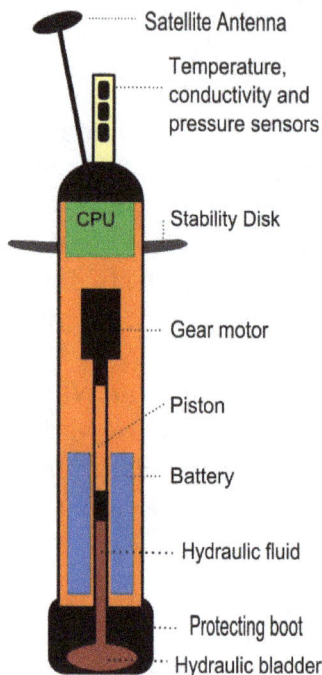

Satellite Antenna

Temperature, conductivity and pressure sensors

Stability Disk

CPU

Gear motor

Piston

Battery

Hydraulic fluid

Protecting boot

Hydraulic bladder

A schematic diagram showing the general structure of a profiling float as used in Argo

A handful of companies and organizations manufacture profiling floats used in the Argo program. APEX floats, made by Teledyne Webb Research, are the most common element of the current array. SOLO and SOLO-II floats (the latter use a reciprocating pump for buoyancy changes, unlike screw-driven pistons in other floats) were developed at Scripps Institution of Oceanography. Other types include the NINJA float, made by the Tsurumi Seiki Co. of Japan, and the PROVOR float developed by IFREMER in France. Most floats use sensors made by Sea-Bird Electronics, which also makes a profiling float called Navis. A typical Argo float is a cylinder just over 1 metre long and 14 cm across with a hemispherical cap. Thus it has a minimum volume of about 16,600 cubic centimetres (cm³). At Ocean Station Papa in the Gulf of Alaska the temperature and salinity at the surface might be about 6 °C and 32.55 parts per thousand giving a density of sea-water of 1.0256 g/cm³. At a depth of 2000 metres (pressure of 2000 decibars) the temperature might be 2 °C and the salinity 34.58 parts per thousand. Thus, including the effect of pressure (water is slightly compressible) the density of sea-water is about 1.0369 g/cm³. The change in density divided by the deep density is 0.0109.

The float has to match these densities if it is to reach 2000 metres depth and then rise to the surface. Since the density of the float is its mass divided by volume, it needs to change its volume by 0.0109 × 16,600 = 181 cm³ to drive that excursion; a small amount of that volume change is provided by the compressibility of the float itself, and excess buoyancy is required at the surface in order to keep the antenna above water. All Argo floats carry sensors to measure the temperature and salinity of the ocean as they vary with depth, but an increasing number of floats also carry other sensors, such as for measuring dissolved oxygen and ultimately other variables of biological and chemical interest such as chlorophyll, nutrients and pH. An extension to the Argo project called BioArgo is being developed and, when implemented, will add a biological and chemical component to this method of sampling the oceans.

The antenna for satellite communications is mounted at the top of the float which extends clear of the sea surface after it completes its ascent. The ocean is saline, hence an electrical conductor, so that radio communications from under the sea surface are not possible. Early in the program Argo floats exclusively used slow mono-directional satellite communications but the majority of floats being deployed in mid-2013 use rapid bi-directional communications. The result of this is that Argo floats now transmit much more data than was previously possible and they spend only about 20 minutes on the sea surface rather than 8–12 hours, greatly reducing problems such as grounding and bio-fouling.

The average life span of Argo floats has increased greatly since the program began, first exceeding 4-year mean lifetime for floats deployed in 2005. Ongoing improvements should result in further extensions to 6 years and longer.

As of March 2016, new types of floats were being tested to collect measurements much deeper than can be reached by standard Argo floats. These "Deep Argo" floats are designed to reach depths of 6000 metres, versus 2000 metres for standard floats. This will allow a much greater volume of the ocean to be sampled. Such measurements are important for developing a comprehensive understanding of the ocean, such as trends in heat content.

Array Design

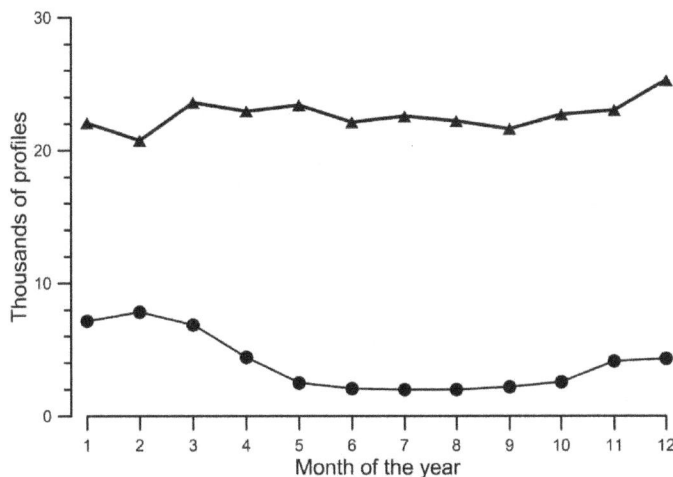

The number of profiles gathered by Argo floats south of 30°S (upper curve) compared with profiles gathered by other means (lower). This shows the near elimination of the seasonal bias.

The original plan advertised in the Argo prospectus called for a nearest-neighbour distance between floats, on average, of 3° latitude by 3° longitude. This allowed for higher resolution (in kilometres) at high latitudes, both north and south, and was considered necessary because of the decrease in the Rossby radius of deformation which governs the scale of oceanographic features, such as eddies. By 2007 this was largely achieved, but the target resolution has never yet been completely achieved in the deep southern ocean.

Efforts are being made to complete the original plan in all parts of the world oceans but this is difficult in the deep Southern Ocean as deployment opportunities occur only very rarely.

As mentioned in the history section, enhancements are now planned in the equatorial regions of the oceans, in boundary currents and in marginal seas. This requires that the total number of floats be increased from the original plan of 3000 floats to a 4000-float array.

One consequence of the use of profiling floats to sample the ocean is that seasonal bias can be removed. The diagram opposite shows the count of all float profiles acquired each month by Argo south of 30°S (upper curve) from the start of the program to November 2012 compared with the same diagram for all other data available. The lower curve shows a strong annual bias with four times as many profiles being collected in austral summer than in austral winter. For the upper (Argo) plot, there is no bias apparent.

Data Access

One of the critical features of the Argo model is that of global and unrestricted access to data in near real-time. When a float transmits a profile it is quickly converted to a format that can be inserted on the GTS (Global Telecommunications System). The GTS is operated by the World Meteorological Organisation, or WMO, specifically for the purpose of sharing data needed for weather forecasting. Thus all nations who are members of the WMO receive all Argo profiles within a few hours of the acquisition of the profile. Data are also made available through ftp and WWW access via two Argo Global Data Centres (or GDACs), one in France and one in the US.

About 90% of all profiles acquired are made available to global access within 24 hours, with the remaining profiles becoming available soon thereafter.

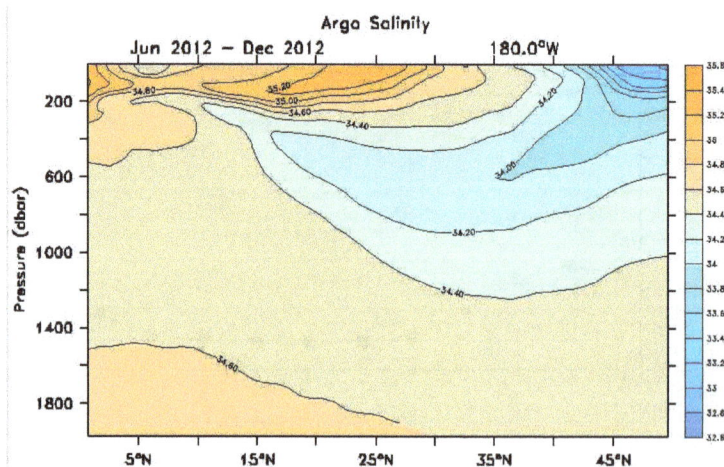

An actual section of salinity along the date line computed from Argo data using the Global Marine Atlas.

For a researcher to use data acquired via the GTS or from the Argo Global Data Centres (GDACs) does require programming skills. The GDACs supply multi-profile files that are a native file format for Ocean DataView. For any day there are files with names like 20121106_prof.nc that are called multi-profile files. This example is a file specific to 6 November 2012 and contains all profiles in a single NetCDF file for one ocean basin. The GDACs identify three ocean basins, Atlantic, Indian and Pacific. Thus three multi-profile files will carry every Argo profile acquired on that specific day.

A user who wants to explore Argo data but lacks programming skills might like to download the Argo Global Marine Atlas which is an easy-to-use utility that allows the creation of products based on Argo data such as the salinity section shown above, but also horizontal maps of ocean properties, time series at any location etc. This Atlas also carries an "update" button that allows data to be updated periodically. The Argo Global Marine Atlas is maintained at the Scripps Institution of Oceanography in La Jolla, California.

Argo data can also be displayed in Google Earth with a layer developed by the Argo Technical Coordinator.

Data Results

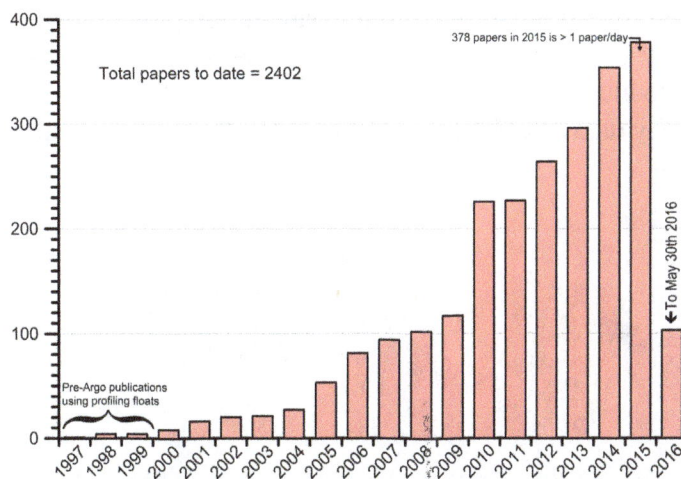

The number of papers, by year, published in refereed journals and that are extensively or totally dependent on the availability of Argo data as of 30th May 2016.

Argo is now the dominant source of information about the climatic state of the oceans and is being widely used in many publications as seen in the diagram opposite. Topics addressed include air-sea interaction, ocean currents, interannual variability, El Niño, mesoscale eddies, water mass properties and transformation. Argo is also now permitting direct computations of the global ocean heat content.

A notable recent paper was published by Durack and Wijffels and analyses global changes in surface salinity patterns.

They determine that areas of the world with high surface salinity are getting saltier and areas of the world with relatively low surface salinity are getting fresher. This has been described as 'the rich get richer and the poor get poorer'. Scientifically speaking, the distributions of salt are governed by the difference between precipitation and evaporation. Areas, such as the northern North Pacific Ocean, where precipitation dominates evaporation are fresher than average. The implication of their result is that the Earth is seeing an intensification of the global hydrological cycle. Argo data are also being used to drive computer models of the climate system leading to improvements in the ability of nations to forecast seasonal climate variations.

Argo data were critical in the drafting of Chapter 3 (Working Group 1) of the IPCC Fifth Assessment Report (released September 2013) and an appendix was added to that chapter to emphasize the profound change that had taken place in the quality and volume of ocean data since the IPCC Fourth Assessment Report and the resulting improvement in confidence in the description of surface salinity changes and upper-ocean heat content.

Bathymetry

Bathymetry is the study of underwater depth of lake or oc ean floors. In other words, bathymetry is the underwater equivalent to hypsometry or topography. Bathymetric (or hydrographic) charts are typically produced to support safety of surface or sub-surface navigation, and usually show seafloor relief or terrain as contour lines (called depth contours or isobaths) and selected depths (soundings), and typically also provide surface navigational information. Bathymetric maps (a more general term where navigational safety is not a concern) may also use a Digital Terrain Model and artificial illumination techniques to illustrate the depths being portrayed. The global bathymetry is sometimes combined with topography data to yield a Global Relief Model. Paleobathymetry is the study of past underwater depths.

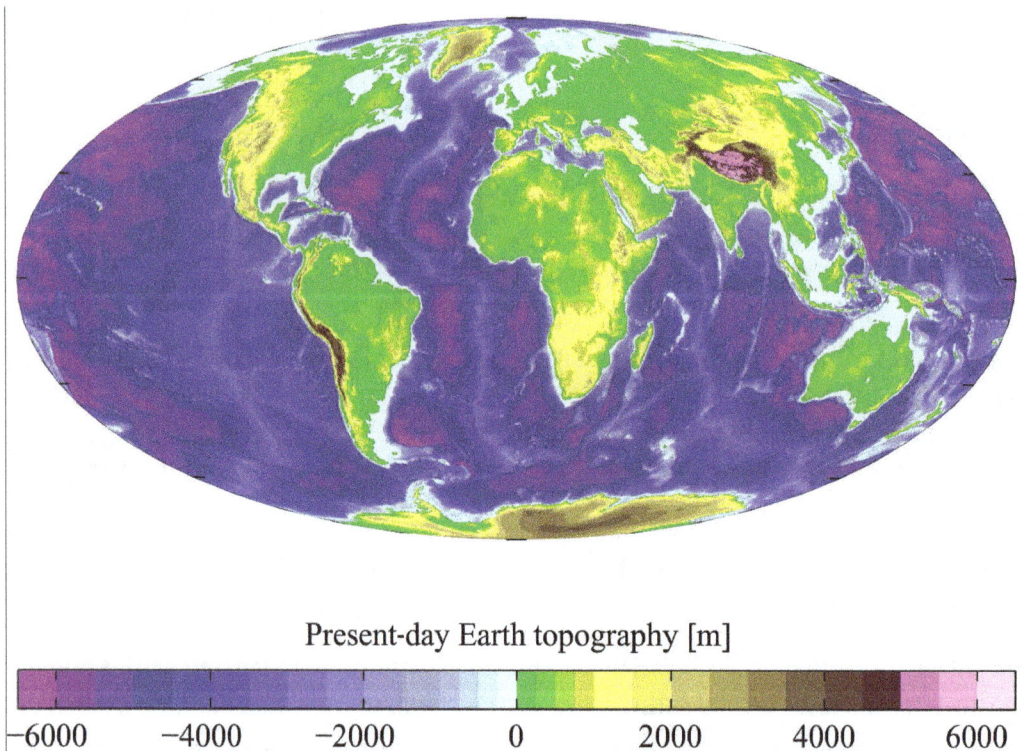

Present-day Earth topography [m]

−6000   −4000   −2000   0   2000   4000   6000

Present day Earth bathymetry (and altimetry). Data from the National Geophysical Data Center's TerrainBase Digital Terrain Model.

Measurement

First printed map of oceanic bathymetry, produced with data from USS *Dolphin* (1853).

Originally, bathymetry involved the measurement of ocean depth through depth sounding. Early techniques used pre-measured heavy rope or cable lowered over a ship's side. This technique measures the depth only a singular point at a time, and is therefore inefficient. It is also subject to movements of the ship and currents moving the line out of true and therefore is inaccurate.

The data used to make bathymetric maps today typically comes from an echosounder (sonar) mounted beneath or over the side of a boat, "pinging" a beam of sound downward at the seafloor or from remote sensing LIDAR or LADAR systems. The amount of time it takes for the sound or light to travel through the water, bounce off the seafloor, and return to the sounder informs the equipment of the distance to the seafloor. LIDAR/LADAR surveys are usually conducted by airborne systems.

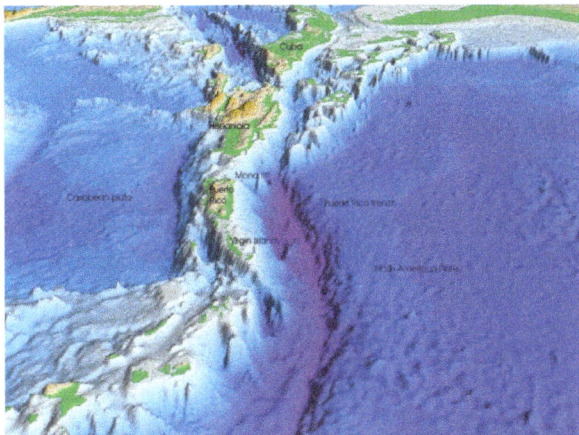

The seafloor topography near the Puerto Rico Trench

Starting in the early 1930s, single-beam sounders were used to make bathymetry maps. Today, multibeam echosounders (MBES) are typically used, which use hundreds of very narrow adjacent beams arranged in a fan-like swath of typically 90 to 170 degrees across. The tightly packed array of narrow individual beams provides very high angular resolution and accuracy. In general a wide swath, which is depth dependent, allows a boat to map more seafloor in less time than a single-beam echosounder by making fewer passes. The beams update many times per second (typically 0.1-50 Hz depending on water depth), allowing faster boat speed while maintaining 100%

coverage of the seafloor. Attitude sensors allow for the correction of the boat's roll and pitch on the ocean surface, and a gyrocompass provides accurate heading information to correct for vessel yaw. (Most modern MBES systems use an integrated motion-sensor and position system that measures yaw as well as the other dynamics and position.) A boat-mounted Global Positioning System (GPS) (or other Global Navigation Satellite System (GNSS)) positions the soundings with respect to the surface of the earth. Sound speed profiles (speed of sound in water as a function of depth) of the water column correct for refraction or "ray-bending" of the sound waves owing to non-uniform water column characteristics such as temperature, conductivity, and pressure. A computer system processes all the data, correcting for all of the above factors as well as for the angle of each individual beam. The resulting sounding measurements are then processed either manually, semi-automatically or automatically (in limited circumstances) to produce a map of the area. As of 2010 a number of different outputs are generated, including a sub-set of the original measurements that satisfy some conditions (e.g., most representative likely soundings, shallowest in a region, etc.) or integrated Digital Terrain Models (DTM) (e.g., a regular or irregular grid of points connected into a surface). Historically, selection of measurements was more common in hydrographic applications while DTM construction was used for engineering surveys, geology, flow modeling, etc. Since ca. 2003-2005, DTMs have become more accepted in hydrographic practice.

Satellites are also used to measure bathymetry. Satellite radar maps deep-sea topography by detecting the subtle variations in sea level caused by the gravitational pull of undersea mountains, ridges, and other masses. On average, sea level is higher over mountains and ridges than over abyssal plains and trenches.

In the United States the United States Army Corps of Engineers performs or commissions most surveys of navigable inland waterways, while the National Oceanic and Atmospheric Administration (NOAA) performs the same role for ocean waterways. Coastal bathymetry data is available from NOAA's National Geophysical Data Center (NGDC), which is now merged into National Centers for Environmental Information. Bathymetric data is usually referenced to tidal vertical datums. For deep-water bathymetry, this is typically Mean Sea Level (MSL), but most data used for nautical charting is referenced to Mean Lower Low Water (MLLW) in American surveys, and Lowest Astronomical Tide (LAT) in other countries. Many other datums are used in practice, depending on the locality and tidal regime.

Occupations or careers related to bathymetry include the study of oceans and rocks and minerals on the ocean floor, and the study of underwater earthquakes or volcanoes. The taking and analysis of bathymetric measurements is one of the core areas of modern hydrography, and a fundamental component in ensuring the safe transport of goods worldwide.

Echo Sounding

Echo sounding is a type of SONAR used to determine the depth of water by transmitting sound pulses into water. The time interval between emission and return of a pulse is recorded, which is used to determine the depth of water along with the speed of sound in water at the time. This information is then typically used for navigation purposes or in order to obtain depths for charting purposes. Echo sounding can also refer to hydroacoustic "echo sounders" defined as active sound in water (sonar) used to study fish. Hydroacoustic assessments have traditionally employed mobile surveys from boats to evaluate fish biomass and spatial distributions. Conversely, fixed-location techniques use stationary transducers to monitor passing fish.

The word *sounding* is used for all types of depth measurements, including those that don't use sound, and is unrelated in origin to the word *sound* in the sense of noise or tones. Echo sounding is a more rapid method of measuring depth than the previous technique of lowering a sounding line until it touched bottom.

Technique

Distance is measured by multiplying half the time from the signal's outgoing pulse to its return by the speed of sound in the water, which is approximately 1.5 kilometres per second [T÷2×(4700 feet per second or 1.5 kil per second)] For precise applications of echosounding, such as hydrography, the speed of sound must also be measured typically by deploying a sound velocity probe into the water. Echo sounding is effectively a special purpose application of sonar used to locate the bottom. Since a traditional pre-SI unit of water depth was the fathom, an instrument used for determining water depth is sometimes called a *fathometer*. The first practical fathometer was invented by Herbert Grove Dorsey and patented in 1928.

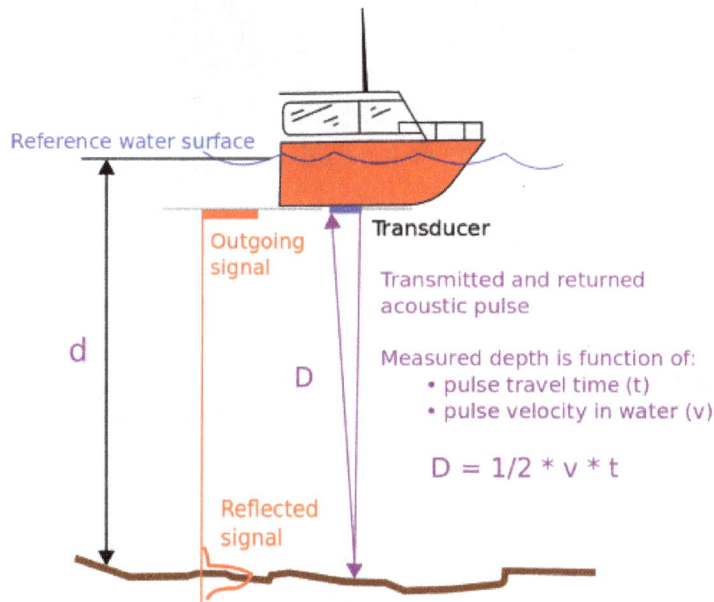

Figure 9-1. Acoustic depth measurement

Diagram showing the basic principle of echo sounding

Most charted ocean depths use an average or standard sound speed. Where greater accuracy is required average and even seasonal standards may be applied to ocean regions. For high accuracy depths, usually restricted to special purpose or scientific surveys, a sensor may be lowered to measure the temperature, pressure and salinity. These factors are used to calculate the actual sound speed in the local water column. This latter technique is regularly used by US Office of Coast Survey for navigational surveys of US coastal waters.

Common Use

As well as an aid to navigation (most larger vessels will have at least a simple depth sounder), echo sounding is commonly used for fishing. Variations in elevation often represent places where fish congregate. Schools of fish will also register. A fishfinder is an echo sounding device used by both recreational and commercial fishers.

Hydrography

In areas where detailed bathymetry is required, a precise echo sounder may be used for the work of hydrography. There are many considerations when evaluating such a system, not limited to the vertical accuracy, resolution, acoustic beamwidth of the transmit/receive beam and the acoustic frequency of the transducer.

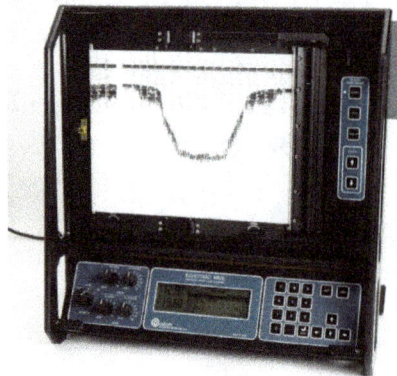

An example of a precision dual frequency echosounder, the Teledyne Odom MkIII

The majority of hydrographic echosounders are dual frequency, meaning that a low frequency pulse (typically around 24 kHz) can be transmitted at the same time as a high frequency pulse (typically around 200 kHz). As the two frequencies are discrete, the two return signals do not typically interfere with each other. There are many advantages of dual frequency echosounding, including the ability to identify a vegetation layer or a layer of soft mud on top of a layer of rock.

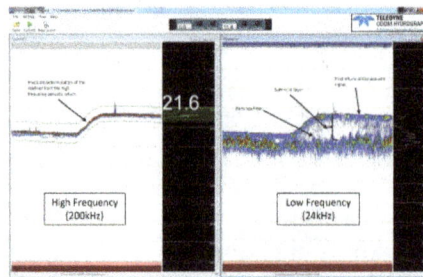

A screen grab of the difference between single and dual frequency echograms

Most hydrographic operations use a 200 kHz transducer, which is suitable for inshore work up to 100 metres in depth. Deeper water requires a lower frequency transducer as the acoustic signal of lower frequencies is less susceptible to attenuation in the water column. Commonly used frequencies for deep water sounding are 33 kHz and 24 kHz.

The beamwidth of the transducer is also a consideration for the hydrographer, as to obtain the best resolution of the data gathered a narrow beamwidth is preferable. This is especially important when sounding in deep water, as the resulting footprint of the acoustic pulse can be very large once it reaches a distant sea floor.

In addition to the single beam echo sounder, there are echo sounders that are capable of receiving many return "pings". These systems are detailed further in the section called multibeam echosounder.

Echo sounders are used in laboratory applications to monitor sediment transport, scour and erosion processes in scale models (hydraulic models, flumes etc.). These can also be used to create plots of 3D contours.

Standards for Hydrographic Echo Sounding

The required precision and accuracy of the hydrographic echo sounder is defined by the requirements of the International Hydrographic Organization (IHO) for surveys that are to be undertaken to IHO standards. These values are contained within IHO publication S44.

In order to meet these standards, the surveyor must consider not only the vertical and horizontal accuracy of the echo sounder and transducer, but the survey system as a whole. A motion sensor may be used, specifically the heave component (in single beam echosounding) to reduce soundings for the motion of the vessel experienced on the water's surface. Once all of the uncertainties of each sensor are established, the hydrographer will create an uncertainty budget to determine whether the survey system meets the requirements laid down by IHO.

Different hydrographic organisations will have their own set of field procedures and manuals to guide their surveyors to meet the required standards. Two examples are the US Army Corps of Engineers publication EM110-2-1003, and the NOAA 'Field Procedures Manual'.

History

German inventor Alexander Behm was granted German patent No. 282009 for the invention of echo sounding (*device for measuring depths of the sea and distances and headings of ships or obstacles by means of reflected sound waves*) on 22 July 1913.

Norwegian Inventor Hans Sundt Berggraf (1874-1941) published the same invention 8 years earlier, 8th of September 1904 in Teknisk Ukeblad.

Ocean Acoustic Tomography

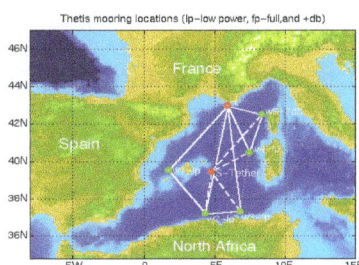

Thetis mooring locations (lp–low power, fp–full, and +db)

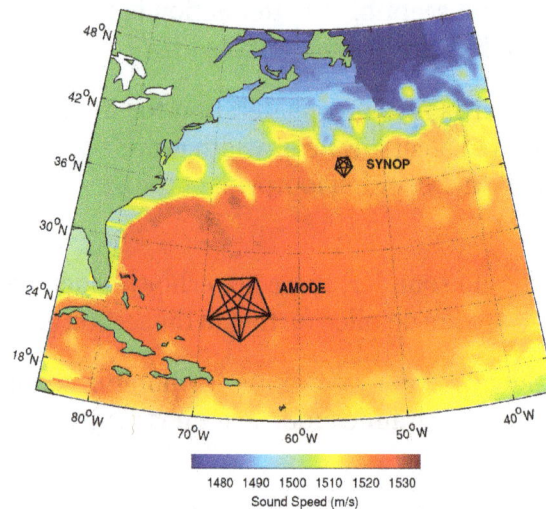

The western North Atlantic showing the locations of two experiments that employed ocean acoustic tomography. AMODE, the "Acoustic Mid-Ocean Dynamics Experiment" (1990-1), was designed to study ocean dynamics in an area away from the Gulf Stream, and SYNOP (1988-9) was designed to synoptically measure aspects of the Gulf Stream. The colors show a snapshot of sound speed at 300 m depth derived from a high-resolution numerical ocean model. One of the key motivations for employing tomography is that the measurements give averages over the turbulent ocean.

Ocean Acoustic Tomography is a technique used to measure temperatures and currents over large regions of the ocean. On ocean basin scales, this technique is also known as acoustic thermometry. The technique relies on precisely measuring the time it takes sound signals to travel between two instruments, one an acoustic source and one a receiver, separated by ranges of 100–5000 km. If the locations of the instruments are known precisely, the measurement of time-of-flight can be used to infer the speed of sound, averaged over the acoustic path. Changes in the speed of sound are primarily caused by changes in the temperature of the ocean, hence the measurement of the travel times is equivalent to a measurement of temperature. A 1 °C change in temperature corresponds to about 4 m/s change in sound speed. An oceanographic experiment employing tomography typically uses several source-receiver pairs in a moored array that measures an area of ocean.

Motivation

Seawater is an electrical conductor, so the oceans are opaque to electromagnetic energy (e.g., light or radar). The oceans are fairly transparent to low-frequency acoustics, however. The oceans conduct sound very efficiently, particularly sound at low frequencies, i.e., less than a few hundred hertz. These properties motivated Walter Munk and Carl Wunsch to suggest "acoustic tomography" for ocean measurement in the late 1970s. The advantages of the acoustical approach to measuring temperature are twofold. First, large areas of the ocean's interior can be measured by remote sensing. Second, the technique naturally averages over the small scale fluctuations of temperature (i.e., noise) that dominate ocean variability.

From its beginning, the idea of observations of the ocean by acoustics was married to estimation of the ocean's state using modern numerical ocean models and the techniques assimilating data into numerical models. As the observational technique has matured, so too have the methods of data assimilation and the computing power required to perform those calculations.

Multipath Arrivals and Tomography

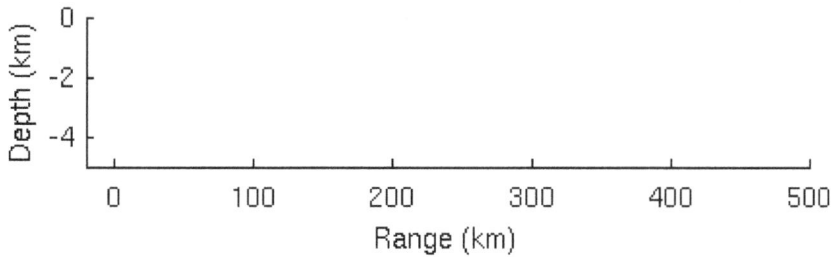

Propagation of acoustic ray paths through the ocean. From the acoustic source at left, the paths are refracted by faster sound speed above and below the SOFAR channel, hence they oscillate about the channel axis. Tomography exploits these "multipaths" to infer information about temperature variations as a function of depth. Note that the aspect ratio of the figure has been greatly skewed to better illustrate the rays; the maximum depth of the figure is only 4.5 km, while the maximum range is 500 km.

One of the intriguing aspects of tomography is that it exploits the fact that acoustic signals travel along a set of generally stable ray paths. From a single transmitted acoustic signal, this set of rays gives rise to multiple arrivals at the receiver, the travel time of each arrival corresponding to a particular ray path. The earliest arrivals correspond to the deeper-traveling rays, since these rays travel where sound speed is greatest. The ray paths are easily calculated using computers ("ray tracing"), and each ray path can generally be identified with a particular travel time. The multiple travel times measure the sound speed averaged over each of the multiple acoustic paths. These measurements make it possible to infer aspects of the structure of temperature or current variations as a function of depth. The solution for sound speed, hence temperature, from the acoustic travel times is an inverse problem.

The Integrating Property of Long-range Acoustic Measurements

Ocean acoustic tomography integrates temperature variations over large distances, that is, the measured travel times result from the accumulated effects of all the temperature variations along the acoustic path, hence measurements by the technique are inherently averaging. This is an important, unique property, since the ubiquitous small-scale turbulent and internal-wave features of the ocean usually dominate the signals in measurements at single points. For example, measurements by thermometers (i.e., moored thermistors or Argo drifting floats) have to contend with this 1-2 °C noise, so that large numbers of instruments are required to obtain an accurate measure of average temperature. For measuring the average temperature of ocean basins, therefore, the acoustic measurement is quite cost effective. Tomographic measurements also average variability over depth as well, since the ray paths cycle throughout the water column.

Reciprocal Tomography

"Reciprocal tomography" employs the simultaneous transmissions between two acoustic transceivers. A "transceiver" is an instrument incorporating both an acoustic source and a receiver. The slight differences in travel time between the reciprocally-traveling signals are used to measure ocean currents, since the reciprocal signals travel with and against the current. The average of these reciprocal travel times is the measure of temperature, with the small effects from ocean currents entirely removed. Ocean temperatures are inferred from the *sum* of reciprocal travel times, while the currents are inferred from the *difference* of reciprocal travel times. Generally, ocean currents (typically 10 cm/s) have a much smaller effect on travel times than sound speed variations (typically 5 m/s), so "one-way" tomography measures temperature to good approximation.

Applications

In the ocean, large-scale temperature changes can occur over time intervals from minutes (internal waves) to decades (oceanic climate change). Tomography has been employed to measure variability over this wide range of temporal scales and over a wide range of spatial scales. Indeed, tomography has been contemplated as a measurement of ocean climate using transmissions over antipodal distances.

Tomography has come to be a valuable method of ocean observation, exploiting the characteristics of long-range acoustic propagation to obtain synoptico. measurements of average ocean temperature or current. One of the earliest applications of tomography in ocean observation occurred in 1988-9. A collaboration between groups at the Scripps Institution of Oceanography and the Woods Hole Oceanographic Institution deployed a six-element tomographic array in the abyssal plain of the Greenland Sea gyre to study deep water formation and the gyre circulation. Other applications include the measurement of ocean tides, and the estimation of ocean mesoscale dynamics by combining tomography, satellite altimetry, and in situ data with ocean dynamical models. In addition to the decade-long measurements obtained in the North Pacific, acoustic thermometry has been employed to measure temperature changes of the upper layers of the Arctic ocean basins, which continues to be an area of active interest. Acoustic thermometry was also recently been used to determine changes to global-scale ocean temperatures using data from acoustic pulses sent from one end of the earth to the other.

Acoustic Thermometry

Acoustic thermometry is an idea to observe the world's ocean basins, and the ocean climate in particular, using trans-basin acoustic transmissions. "Thermometry", rather than "tomography", has been used to indicate basin-scale or global scale measurements. Prototype measurements of temperature have been made in the North Pacific Basin and across the Arctic Basin.

Starting in 1983, John Spiesberger of the Woods Hole Oceanographic Institution, and Ted Birdsall and Kurt Metzger of the University of Michigan developed the use of sound to infer information about the ocean's large-scale temperatures, and in particular to attempt the detection of global warming in the ocean. This group transmitted sounds from Oahu that were recorded at about ten receivers stationed around the rim of the Pacific Ocean over distances of 4000 km. These experiments demonstrated that changes in temperature could be measured with an accuracy of about 20 millidegrees. Spiesberger et al. did not detect global warming. Instead they discovered that other natural climatic fluctuations, such as El Nino, were responsible in part for substanstial fluctuations in temperature that may have masked any slower and smaller trends that may have occurred from global warming

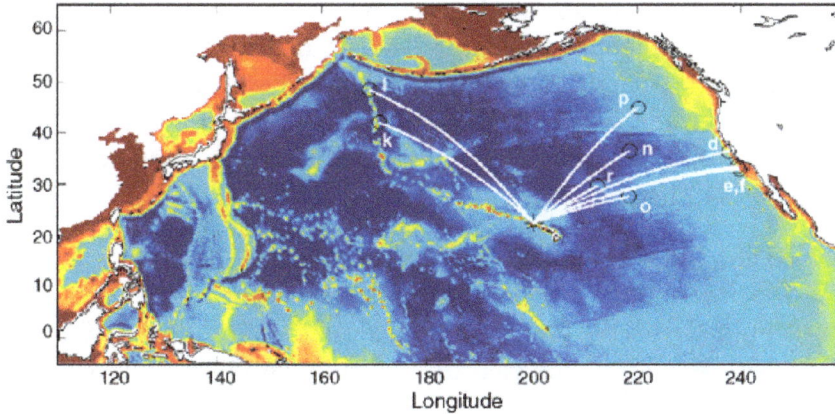

The ATOC prototype array was an acoustic source located just north of Kauai, Hawaii, and transmissions were made to receivers of opportunity in the North Pacific Basin. The source signals were broadband with frequencies centered on 75 Hz and a source level of 195 dB re 1 micropascal at 1 m, or about 250 watts. Six transmissions of 20-minute duration were made on every fourth day.

The Acoustic Thermometry of Ocean Climate (ATOC) program was implemented in the North Pacific Ocean, with acoustic transmissions from 1996 through fall 2006. The measurements terminated when agreed-upon environmental protocols ended. The decade-long deployment of the acoustic source showed that the observations are sustainable on even a modest budget. The transmissions have been verified to provide an accurate measurement of ocean temperature on the acoustic paths, with uncertainties that are far smaller than any other approach to ocean temperature measurement.

Acoustic Transmissions and Marine Mammals

The ATOC project was embroiled in issues concerning the effects of acoustics on marine mammals (e.g. whales, porpoises, sea lions, etc.). Public discussion was complicated by technical issues from a variety of disciplines (physical oceanography, acoustics, marine mammal biology, etc.) that makes understanding the effects of acoustics on marine mammals difficult for the experts, let alone the general public. Many of the issues concerning acoustics in the ocean and their effects on marine mammals were unknown. Finally, there were a variety of public misconceptions initially, such as a confusion of the definition of sound levels in air vs. sound levels in water. If a given number of decibels in water are interpreted as decibels in air, the sound level will seem to be orders of magnitude larger than it really is - at one point the ATOC sound levels were erroneously interpreted as "louder than 10,000 747 airplanes". In fact, the sound powers employed, 250 W, were comparable those made by blue or fin whales, although those whales vocalize at much lower frequencies. The ocean carries sound so efficiently that sounds do not have to be that loud to cross ocean basins. Other factors in the controversy were the extensive history of activism where marine mammals are concerned, stemming from the ongoing whaling conflict, and the sympathy that much of the public feels toward marine mammals.

As a result of this controversy, the ATOC program conducted a $6 million study of the effects of the acoustic transmissions on a variety of marine mammals. After six years of study the official, formal conclusion from this study was that the ATOC transmissions have "no significant biological impact".

Other acoustics activities in the ocean may not be so benign insofar as marine mammals are concerned. Various types of man-made sounds have been studied as potential threats to marine mammals, such as airgun shots for geophysical surveys, or transmissions by the U.S. Navy for various purposes. The actual threat depends on a variety of factors beyond noise levels: sound frequency, frequency and duration of transmissions, the nature of the acoustic signal (e.g., a sudden pulse, or coded sequence), depth of the sound source, directionality of the sound source, water depth and local topography, reverberation, etc.

In the case of the ATOC, the source was mounted on the bottom about a half mile deep, hence marine mammals, which are bound to the surface, were generally further than a half mile from the source. This fact, combined with the modest source level, the infrequent 2% duty cycle (the sound is on only 2% of the day), and other such factors, made the sound transmissions benign in its effect on marine life.

Types of Transmitted Acoustic Signals

Tomographic transmissions consist of long coded signals (e.g., "m-sequences") lasting 30 seconds or more. The frequencies employed range from 50 to 1000 Hz and source powers range from 100 to 250 W, depending on the particular goals of the measurements. With precise timing such as from GPS, travel times can be measured to a nominal accuracy of 1 millisecond. While these transmissions are audible near the source, beyond a range of several kilometers the signals are usually below ambient noise levels, requiring sophisticated spread-spectrum signal processing techniques to recover them.

Thermohaline Circulation

Thermohaline circulation (THC) is a part of the large-scale ocean circulation that is driven by global density gradients created by surface heat and freshwater fluxes. The adjective thermohaline derives from *thermo-* referring to temperature and *-haline* referring to salt content, factors which together determine the density of sea water. Wind-driven surface currents (such as the Gulf Stream) travel polewards from the equatorial Atlantic Ocean, cooling en route, and eventually sinking at high latitudes (forming North Atlantic Deep Water). This dense water then flows into the ocean basins. While the bulk of it upwells in the Southern Ocean, the oldest waters (with a transit time of around 1000 years) upwell in the North Pacific. Extensive mixing therefore takes place between the ocean basins, reducing differences between them and making the Earth's oceans a global system. On their journey, the water masses transport both energy (in the form of heat) and matter (solids, dissolved substances and gases) around the globe. As such, the state of the circulation has a large impact on the climate of the Earth.

The thermohaline circulation is sometimes called the ocean conveyor belt, the great ocean conveyor, or the global conveyor belt. On occasion, it is used to refer to the meridional overturning circulation (often abbreviated as MOC). The term MOC, indeed, is more accurate and well defined, as it is difficult to separate the part of the circulation which is actually driven by temperature and salinity alone as opposed to other factors such as the wind and tidal forces. Moreover, temperature and salinity gradients can also lead to circulation effects that are not included in the MOC itself.

Overview

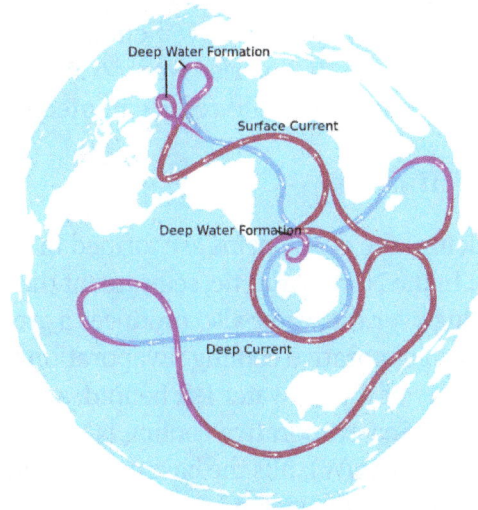

The global conveyor belt on a continuous-ocean map

The movement of surface currents pushed by the wind is fairly intuitive. For example, the wind easily produces ripples on the surface of a pond. Thus the deep ocean — devoid of wind — was assumed to be perfectly static by early oceanographers. However, modern instrumentation shows that current velocities in deep water masses can be significant (although much less than surface speeds).

In the deep ocean, the predominant driving force is differences in density, caused by salinity and temperature variations (increasing salinity and lowering the temperature of a fluid both increase its density). There is often confusion over the components of the circulation that are wind and density driven. Note that ocean currents due to tides are also significant in many places; most prominent in relatively shallow coastal areas, tidal currents can also be significant in the deep ocean.

The density of ocean water is not globally homogeneous, but varies significantly and discretely. Sharply defined boundaries exist between water masses which form at the surface, and subsequently maintain their own identity within the ocean. They position themselves one above or below each other according to their density, which depends on both temperature and salinity.

Warm seawater expands and is thus less dense than cooler seawater. Saltier water is denser than fresher water because the dissolved salts fill interstices between water molecules, resulting in more mass per unit volume. Lighter water masses float over denser ones (just as a piece of wood or ice will float on water). This is known as "stable stratification". When dense water masses are first formed, they are not stably stratified. In order to take up their most stable positions, water masses of different densities must flow, providing a driving force for deep currents.

The thermohaline circulation is mainly triggered by the formation of deep water masses in the North Atlantic and the Southern Ocean caused by differences in temperature and salinity of the water.

The great quantities of dense water sinking at polar ocean basin edges must be offset by equal quantities of water rising elsewhere. Note that cold water in polar zones sink relatively rapidly over a small area, while warm water in temperate and tropical zones rise more gradually across a much larger area. It then slowly returns poleward near the surface to repeat the cycle. The continual diffuse upwelling of deep water maintains the existence of the permanent thermocline found everywhere at low and mid-latitudes. This slow upward movement is estimated to be about 1 centimeter (0.5 inch) per day over most of the ocean. If this rise were to stop, downward movement of heat would cause the thermocline to descend and would reduce its steepness.

Formation of Deep Water Masses

The dense water masses that sink into the deep basins are formed in quite specific areas of the North Atlantic and the Southern Ocean. In the North Atlantic, seawater at the surface of the ocean is intensely cooled by the wind. Wind moving over the water also produces a great deal of evaporation, leading to a decrease in temperature, called evaporative cooling. Evaporation removes only water molecules, resulting in an increase in the salinity of the seawater left behind, and thus an increase in the density of the water mass. In the Norwegian Sea evaporative cooling is predominant, and the sinking water mass, the North Atlantic Deep Water (NADW), fills the basin and spills southwards through crevasses in the submarine sills that connect Greenland, Iceland and Great Britain. It then flows very slowly into the deep abyssal plains of the Atlantic, always in a southerly direction. Flow from the Arctic Ocean Basin into the Pacific, however, is blocked by the narrow shallows of the Bering Strait.

In the Southern Ocean, strong katabatic winds blowing from the Antarctic continent onto the ice shelves will blow the newly formed sea ice away, opening polynyas along the coast. The ocean, no longer protected by sea ice, suffers a brutal and strong cooling. Meanwhile, sea ice starts reforming, so the surface waters also get saltier, hence very dense. In fact, the formation of sea ice contributes to an increase in surface seawater salinity; saltier brine is left behind as the sea ice forms around it (pure water preferentially being frozen). Increasing salinity lowers the freezing point of seawater, so cold liquid brine is formed in inclusions within a honeycomb of ice. The brine progressively melts the ice just beneath it, eventually dripping out of the ice matrix and sinking. This process is known as brine rejection.

The resulting Antarctic Bottom Water (AABW) sinks and flows north and east, but is so dense it actually underflows the NADW. AABW formed in the Weddell Sea will mainly fill the Atlantic and Indian Basins, whereas the AABW formed in the Ross Sea will flow towards the Pacific Ocean.

The dense water masses formed by these processes flow downhill at the bottom of the ocean, like a stream within the surrounding less dense fluid, and fill up the basins of the polar seas. Just as river valleys direct streams and rivers on the continents, the bottom topography steers the deep and bottom water masses.

Note that, unlike fresh water, seawater does not have a density maximum at 4 °C but gets denser as it cools all the way to its freezing point of approximately -1.8 °C.

Movement of Deep Water Masses

Formation and movement of the deep water masses at the North Atlantic Ocean, creates sinking water masses that fill the basin and flows very slowly into the deep abyssal plains of the Atlantic. This high-latitude cooling and the low-latitude heating drives the movement of the deep water in a polar southward flow. The deep water flows through the Antarctic Ocean Basin around South Africa where it is split into two routes: one into the Indian Ocean and one past Australia into the Pacific.

At the Indian Ocean, some of the cold and salty water from the Atlantic — drawn by the flow of warmer and fresher upper ocean water from the tropical Pacific — causes a vertical exchange of dense, sinking water with lighter water above. It is known as overturning. In the Pacific Ocean, the rest of the cold and salty water from the Atlantic undergoes haline forcing, and becomes warmer and fresher more quickly.

The out-flowing undersea of cold and salty water makes the sea level of the Atlantic slightly lower than the Pacific and salinity or halinity of water at the Atlantic higher than the Pacific. This generates a large but slow flow of warmer and fresher upper ocean water from the tropical Pacific to the Indian Ocean through the Indonesian Archipelago to replace the cold and salty Antarctic Bottom Water. This is also known as 'haline forcing' (net high latitude freshwater gain and low latitude evaporation). This warmer, fresher water from the Pacific flows up through the South Atlantic to Greenland, where it cools off and undergoes evaporative cooling and sinks to the ocean floor, providing a continuous thermohaline circulation.

Hence, a recent and popular name for the thermohaline circulation, emphasizing the vertical nature and pole-to-pole character of this kind of ocean circulation, is the meridional overturning circulation.

Quantitative Estimation

Direct estimates of the strength of the thermohaline circulation have been made at 26.5°N in the North Atlantic since 2004 by the UK-US RAPID programme. By combining direct estimates of ocean transport using current meters and subsea cable measurements with estimates of the geostrophic current from temperature and salinity measurements, the RAPID programme provides continuous, full-depth, basinwide estimates of the thermohaline circulation or, more accurately, the meridional overturning circulation.

The deep water masses that participate in the MOC have chemical, temperature and isotopic ratio signatures and can be traced, their flow rate calculated, and their age determined. These include ^{231}Pa / ^{230}Th ratios.

Gulf Stream

Benjamin Franklin's map of the Gulf Stream

The Gulf Stream, together with its northern extension towards Europe, the North Atlantic Drift, is a powerful, warm, and swift Atlantic ocean current that originates at the tip of Florida, and follows the eastern coastlines of the United States and Newfoundland before crossing the Atlantic Ocean. The process of western intensification causes the Gulf Stream to be a northward accelerating current off the east coast of North America. At about

40°0′N 30°0′W40.000°N 30.000°W, it splits in two, with the northern stream crossing to northern Europe and the southern stream recirculating off West Africa. The Gulf Stream influences the climate of the east coast of North America from Florida to Newfoundland, and the west coast of Europe. Although there has been recent debate, there is consensus that the climate of Western Europe and Northern Europe is warmer than it would otherwise be due to the North Atlantic drift, one of the branches from the tail of the Gulf Stream. It is part of the North Atlantic Gyre. Its presence has led to the development of strong cyclones of all types, both within the atmosphere and within the ocean. The Gulf Stream is also a significant potential source of renewable power generation.

Upwelling

All these dense water masses sinking into the ocean basins displace the older deep water masses which were made less dense by ocean mixing. To maintain a balance water must be rising elsewhere. However, because this thermohaline upwelling is so widespread and diffuse, its speeds are very slow even compared to the movement of the bottom water masses. It is therefore difficult to measure where upwelling occurs using current speeds, given all the other wind-driven processes going on in the surface ocean. Deep waters have their own chemical signature, formed from the breakdown of particulate matter falling into them over the course of their long journey at depth. A number of scientists have tried to use these tracers to infer where the upwelling occurs.

Wallace Broecker, using box models, has asserted that the bulk of deep upwelling occurs in the North Pacific, using as evidence the high values of silicon found in these waters. Other investigators have not found such clear evidence. Computer models of ocean circulation increasingly place most of the deep upwelling in the Southern Ocean, associated with the strong winds in the open latitudes between South America and Antarctica. While this picture is consistent with the global observational synthesis of William Schmitz at Woods Hole and with low observed values of diffusion, not all observational syntheses agree. Recent papers by Lynne Talley at the Scripps Institution of Oceanography and Bernadette Sloyan and Stephen Rintoul in Australia suggest that a significant amount of dense deep water must be transformed to light water somewhere north of the Southern Ocean.

Effects on Global Climate

The thermohaline circulation plays an important role in supplying heat to the polar regions, and thus in regulating the amount of sea ice in these regions, although poleward heat transport outside the tropics is considerably larger in the atmosphere than in the ocean. Changes in the thermohaline circulation are thought to have significant impacts on the Earth's radiation budget. Insofar as the thermohaline circulation governs the rate at which deep waters are exposed to the surface, it may also play an important role in the concentration of carbon dioxide in the atmosphere. While it is often stated that the thermohaline circulation is the primary reason that Western Europe is so temperate, it has been suggested that this is largely incorrect, and that Europe is warm mostly because it lies downwind of an ocean basin, and because of the effect of atmospheric waves bringing warm air north from the subtropics. However, the underlying assumptions of this particular analysis have likewise been challenged.

Large influxes of low-density meltwater from Lake Agassiz and deglaciation in North America are thought to have led to a shifting of deep water formation and subsidence in the extreme North Atlantic and caused the climate period in Europe known as the Younger Dryas.

Shutdown of Thermohaline Circulation

In 2005, British researchers noticed that the net flow of the northern Gulf Stream had decreased by about 30% since 1957. Coincidentally, scientists at Woods Hole had been measuring the freshening of the North Atlantic as Earth becomes warmer. Their findings suggested that precipitation increases in the high northern latitudes, and polar ice melts as a consequence. By flooding the northern seas with lots of extra fresh water, global warming could, in theory, divert the Gulf Stream waters that usually flow northward, past the British Isles and Norway, and cause them to instead circulate toward the equator. If this were to happen, Europe's climate would be seriously impacted.

Downturn of AMOC (Atlantic meridional overturning circulation), has been tied to extreme regional sea level rise.

Ocean: An Overview

An ocean is a body of water, which has significant amounts of dissolved salt in it. Some of the oceans discussed in this chapter are the Artic Ocean, Atlantic Ocean, Pacific and Indian Ocean. It provides examples of and also gives a brief introduction on the deepest part of the world ocean. This chapter is an overview of the subject matter incorporating all the major aspects of ocean.

Ocean

An ocean is a body of saline water that composes much of a planet's hydrosphere. On Earth, an ocean is one of the major conventional divisions of the World Ocean, which covers almost 71% of its surface. These are, in descending order by area, the Pacific, Atlantic, Indian, Southern (Antarctic), and Arctic Oceans. The word *sea* is often used interchangeably with "ocean" in American English but, strictly speaking, a sea is a body of saline water (generally a division of the world ocean) partly or fully enclosed by land.

Saline water covers approximately 72% of the planet's surface ($\sim3.6\times10^8$ km²) and is customarily divided into several principal oceans and smaller seas, with the ocean covering approximately 71% of Earth's surface and 90% of the earth's biosphere. The ocean contains 97% of Earth's water, and oceanographers have stated that less than 5% of the World Ocean has been explored. The total volume is approximately 1.35 billion cubic kilometers (320 million cu mi) with an average depth of nearly 3,700 meters (12,100 ft).

As it is the principal component of Earth's hydrosphere, the world ocean is integral to all known life, forms part of the carbon cycle, and influences climate and weather patterns. It is the habitat of 230,000 known species, although much of the oceans depths remain unexplored, and over two million marine species are estimated to exist. The origin of Earth's oceans remains unknown; oceans are thought to have formed in the Hadean period and may have been the impetus for the emergence of life.

Extraterrestrial oceans may be composed of water or other elements and compounds. The only confirmed large stable bodies of extraterrestrial surface liquids are the lakes of Titan, although there is evidence for the existence of oceans elsewhere in the Solar System. Early in their geologic histories, Mars and Venus are theorized to have had large water oceans. The Mars ocean hypothesis suggests that nearly a third of the surface of Mars was once covered by water, and a runaway greenhouse effect may have boiled away the global ocean of Venus. Compounds such as salts and ammonia dissolved in water lower its freezing point, so that water might exist in large quantities in extraterrestrial environments as brine or convecting ice. Unconfirmed oceans are speculated beneath the surface of many dwarf planets and natural satellites; notably, the ocean of Europa is estimated to have over twice the water volume of Earth. The Solar System's giant planets are also thought to have liquid atmospheric layers of yet to be confirmed compositions. Oceans may also exist on exoplanets and exomoons, including surface oceans of liquid water within a circumstellar habitable zone. Ocean planets are a hypothetical type of planet with a surface completely covered with liquid.

Etymology

The word « ocean » comes from the figure in classical antiquity, the elder of the Titans in classical Greek mythology, believed by the ancient Greeks and Romans to be the divine personification of the sea, an enormous river encircling the world.

Earth's Global Ocean

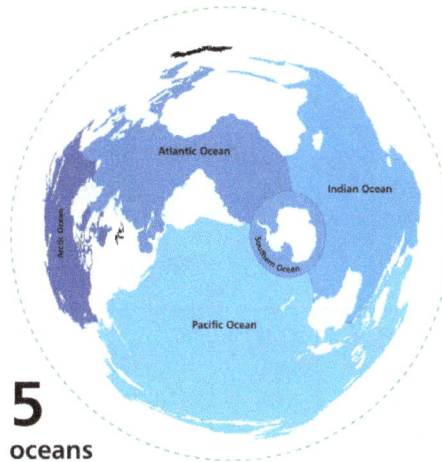

Various ways to divide the World Ocean

Oceanic Divisions

1. Epipelagic zone: surface - 200 meters deep 2. Mesopelagic zone: 200 - 1000m 3. Bathypelagic zone: 1000m - 4000m 4. Abyssopelagic zone: 4000m - 6000m 5. Hadal zone (the trenches): 6000 m to the bottom of the ocean

Though generally described as several separate oceans, these waters comprise one global, interconnected body of salt water sometimes referred to as the World Ocean or global ocean.

This concept of a continuous body of water with relatively free interchange among its parts is of fundamental importance to oceanography.

The major oceanic divisions – listed below in descending order of area and volume – are defined in part by the continents, various archipelagos, and other criteria.

#	Ocean	Location	Area (km²) (%)	Volume (km³) (%)	Avg. depth (m)	Coastline (km)
1	Pacific Ocean	Separates Asia and Oceania from the Americas	168,723,000 46.6	669,880,000 50.1	3,970	135,663
2	Atlantic Ocean	Separates the Americas from Eurasia and Africa	85,133,000 23.5	310,410,900 23.3	3,646	111,866
3	Indian Ocean	Washes upon southern Asia and separates Africa and Australia	70,560,000 19.5	264,000,000 19.8	3,741	66,526
4	Southern Ocean	Sometimes considered an extension of the Pacific, Atlantic and Indian Oceans, which encircles Antarctica	21,960,000 6.1	71,800,000 5.4	3,270	17,968
5	Arctic Ocean	Sometimes considered a sea or estuary of the Atlantic, which covers much of the Arctic and washes upon northern North America and Eurasia	15,558,000 4.3	18,750,000 1.4	1,205	45,389
Total – World Ocean			361,900,000 100	1,335,000,000 100	3,688	377,412

NB: Volume, area, and average depth figures include NOAA ETOPO1 figures for marginal South China Sea. Sources: Encyclopedia of Earth, International Hydrographic Organization, *Regional Oceanography: an Introduction* (Tomczak, 2005), Encyclopædia Britannica, and the International Telecommunication Union.

Oceans are fringed by smaller, adjoining bodies of water such as seas, gulfs, bays, bights, and straits.

Global System

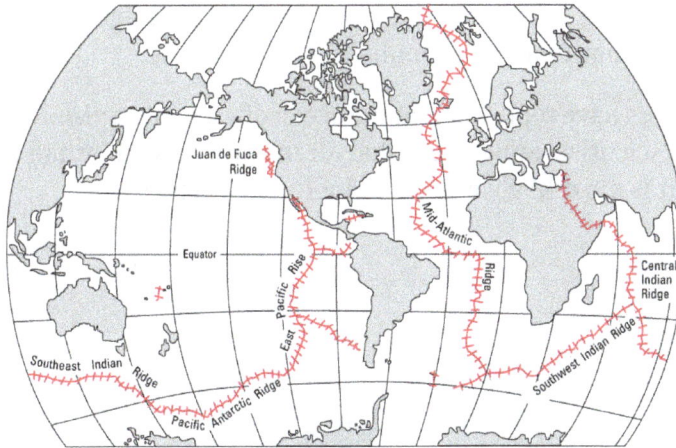

World Distribution of Mid-Oceanic Ridges; USGS

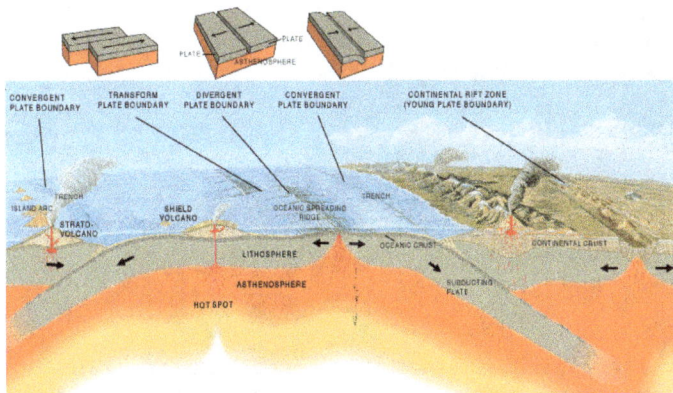

Three main types of plate boundaries.

The Mid-Oceanic Ridge of the World are connected and form *the* Ocean Ridge, a single global mid-oceanic ridge system that is part of every ocean, making it the longest mountain range in the world. The continuous mountain range is 65,000 km (40,400 mi) long (several times longer than the Andes, the longest continental mountain range), and the total length of the oceanic ridge system is 80,000 km (49,700 mi) long.

Physical Properties

The total mass of the hydrosphere is about 1.4 quintillion metric tons (1.4×10^{18} long tons or 1.5×10^{18} short tons), which is about 0.023% of Earth's total mass. Less than 3% is freshwater; the rest is saltwater, almost all of which is in the ocean. The area of the World Ocean is about 361.9 million square kilometers (139.7 million square miles), which covers about 70.9% of Earth's surface, and its volume is approximately 1.335 billion cubic kilometers (320.3 million cubic miles). This can be thought of as a cube of water with an edge length of 1,101 kilometers (684 mi). Its average depth is about 3,688 meters (12,100 ft), and its maximum depth is 10,994 meters (6.831 mi) at the Mariana Trench. Nearly half of the world's marine waters are over 3,000 meters (9,800 ft) deep. The vast expanses of deep ocean (anything below 200 meters or 660 feet) cover about 66% of Earth's surface. This does not include seas not connected to the World Ocean, such as the Caspian Sea.

The bluish color of water is a composite of several contributing agents. Prominent contributors include dissolved organic matter and chlorophyll.

Sailors and other mariners have reported that the ocean often emits a visible glow which extends for miles at night. In 2005, scientists announced that for the first time, they had obtained photographic evidence of this glow. It is most likely caused by bioluminescence.

Oceanic Zones

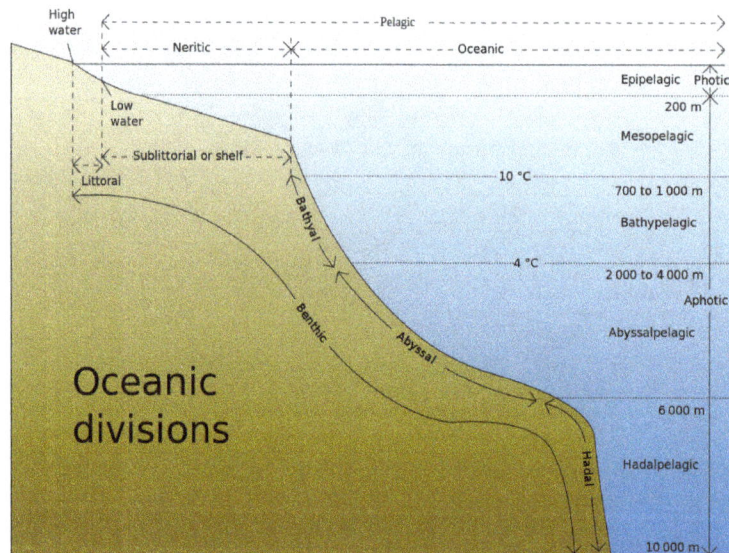

The major oceanic zones, based on depth and biophysical conditions

Oceanographers divide the ocean into different zones by physical and biological conditions. The pelagic zone includes all open ocean regions, and can be divided into further regions categorized by depth and light abundance. The photic zone includes the oceans from the surface to a depth of 200 m; it is the region where photosynthesis can occur and is, therefore, the most biodiverse. Because plants require photosynthesis, life found deeper than the photic zone must either rely on material sinking from above or find another energy source. Hydrothermal vents are the primary source of energy in what is known as the aphotic zone (depths exceeding 200 m). The pelagic part of the photic zone is known as the epipelagic.

The pelagic part of the aphotic zone can be further divided into vertical regions according to temperature. The mesopelagic is the uppermost region. Its lowermost boundary is at a thermocline of 12 °C (54 °F), which, in the tropics generally lies at 700–1,000 meters (2,300–3,300 ft). Next is the bathypelagic lying between 10 and 4 °C (50 and 39 °F), typically between 700–1,000 meters (2,300–3,300 ft) and 2,000–4,000 meters (6,600–13,100 ft) Lying along the top of the abyssal plain is the abyssopelagic, whose lower boundary lies at about 6,000 meters (20,000 ft). The last zone includes the deep oceanic trench, and is known as the hadalpelagic. This lies between 6,000–11,000 meters (20,000–36,000 ft) and is the deepest oceanic zone.

The benthic zones are aphotic and correspond to the three deepest zones of the deep-sea. The bathyal zone covers the continental slope down to about 4,000 meters (13,000 ft). The abyssal zone covers the abyssal plains between 4,000 and 6,000 m. Lastly, the hadal zone corresponds to the hadalpelagic zone, which is found in oceanic trenches.

The pelagic zone can be further subdivided into two subregions: the neritic zone and the oceanic zone. The neritic zone encompasses the water mass directly above the continental shelves whereas the oceanic zone includes all the completely open water.

In contrast, the littoral zone covers the region between low and high tide and represents the transitional area between marine and terrestrial conditions. It is also known as the intertidal zone because it is the area where tide level affects the conditions of the region.

The ocean can be divided into three density zones: the surface zone, the pycnocline, and the deep zone. The surface zone, also called the mixed layer, refers to the uppermost density zone of the ocean. Temperature and salinity are relatively constant with depth in this zone due to currents and wave action. The surface zone contains ocean water that is in contact with the atmosphere and within the photic zone. The surface zone has the ocean's least dense water and represents approximately 2% of the total volume of ocean water. The surface zone usually ranges between depths of 500 feet to 3,300 feet below ocean surface, but this can vary a great deal. In some cases, the surface zone can be entirely non-existent. The surface zone is typically thicker in the tropics than in regions of higher latitude. The transition to colder, denser water is more abrupt in the tropics than in regions of higher latitudes. The pycnocline refers to a zone wherein density substantially increases with depth due primarily to decreases in temperature. The pycnocline effectively separates the lower-density surface zone above from the higher-density deep zone below. The pycnocline represents approximately 18% of the total volume of ocean water. The deep zone refers to the lowermost density zone of the ocean. The deep zone usually begins at depths below 3,300 feet in mid-latitudes. The deep zone undergoes negligible changes in water density with depth. The deep zone represents approximately 80% of the total volume of ocean water. The deep zone contains relatively colder and stable water.

If a zone undergoes dramatic changes in temperature with depth, it contains a thermocline. The tropical thermocline is typically deeper than the thermocline at higher latitudes. Polar waters, which receive relatively little solar energy, are not stratified by temperature and generally lack a thermocline because surface water at polar latitudes are nearly as cold as water at greater depths. Below the thermocline, water is very cold, ranging from –1 °C to 3 °C. Because this deep and cold layer contains the bulk of ocean water, the average temperature of the world ocean is 3.9 °C If a zone undergoes dramatic changes in salinity with depth, it contains a halocline. If a zone undergoes a strong, vertical chemistry gradient with depth, it contains a chemocline.

The halocline often coincides with the thermocline, and the combination produces a pronounced pycnocline.

Exploration

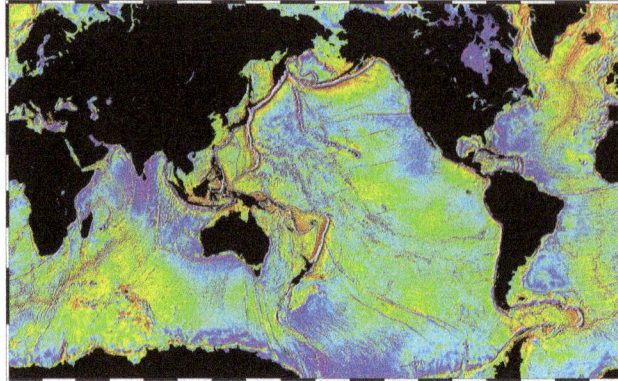

Map of large underwater features (1995, NOAA)

Ocean travel by boat dates back to prehistoric times, but only in modern times has extensive underwater travel become possible.

The deepest point in the ocean is the Mariana Trench, located in the Pacific Ocean near the Northern Mariana Islands. Its maximum depth has been estimated to be 10,971 meters (35,994 ft). The British naval vessel *Challenger II* surveyed the trench in 1951 and named the deepest part of the trench the "Challenger Deep". In 1960, the Trieste successfully reached the bottom of the trench, manned by a crew of two men.

Oceanic Maritime Currents

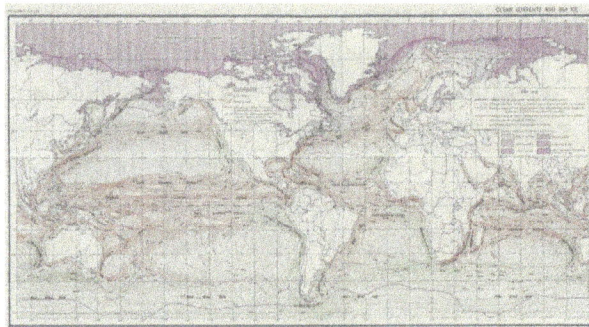

Oceanic currents in 1943.

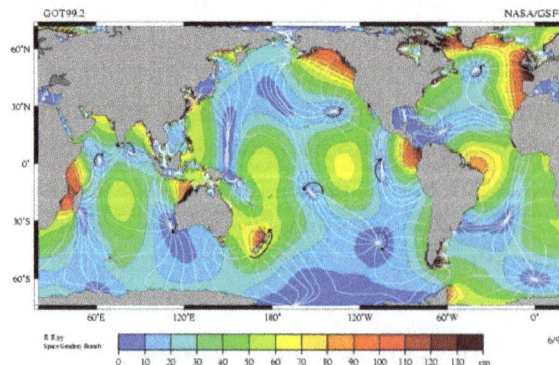

Amphidromic points showing the direction of tides per incrementation periods along with resonating directions of wavelength movements.

Oceanic maritime currents have different origins. Tidal currents are in phase with the tide, hence are quasiperiodic, they may fomulate various knots in certain places, most notably around headlands. Non periodic currents have for origin the waves, wind and different densities.

The wind and waves create surface currents (designated as « drift currents »). These currents can decompose in one quasi permanent current (which varies within the hourly scale) and one movement of Stokes drift under the effect of rapid waves movement (at the echelon of a couple of seconds).). The quasi permanent current is accelerated by the breaking of waves, and in a lesser governing effect, by the friction of the wind on the surface.

This acceleration of the current takes place in the direction of waves and dominant wind. Accordingly, when the sea depth increases, the rotation of the earth changes the direction of currents, in proportion with the increase of depth while friction lowers their speed. At a certain sea depth, the current changes direction and is seen inverted in the opposite direction with speed current becoming nul: known as the Ekman spiral. The influence of these currents is mainly experienced at the mixed layer of the ocean surface, often from 400 to 800 meters of maximum depth. These currents can considerably alter, change and are dependent on the various yearly seasons. If the mixed layer is less thick (10 to 20 meters), the quasi permanent current at the surface adopts an extreme oblique direction in relation to the direction of the wind, becoming virtually homogeneous, until the Thermocline.

In the deep however, maritime currents are caused by the temperature gradients and the salinity between water density masses.

In Littoral zones, Breaking wave is so intense and the depth measurement so low, that maritime currents reach often 1 to 2 knots.

Climate

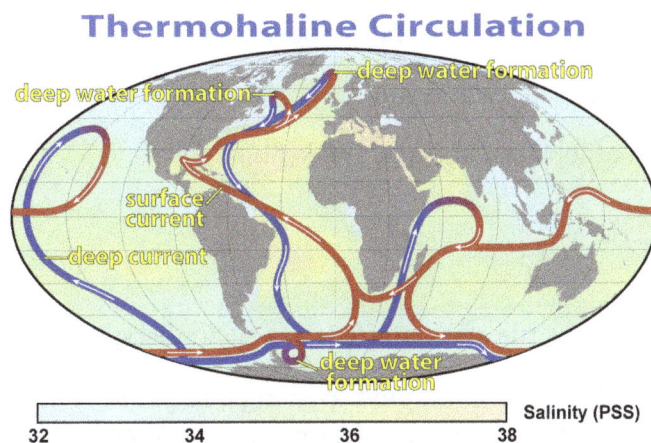

A map of the global thermohaline circulation; blue represent deep-water currents, whereas red represent surface currents

Ocean currents greatly affect Earth's climate by transferring heat from the tropics to the polar regions. Transferring warm or cold air and precipitation to coastal regions, winds may carry them inland. Surface heat and freshwater fluxes create global density gradients that drive the thermohaline circulation part of large-scale ocean circulation. It plays an important role in supplying heat to the polar regions, and thus in sea ice regulation. Changes in the thermohaline circulation are thought to have significant impacts on Earth's energy budget. In so far as the thermohaline circulation governs the rate at which deep waters reach the surface, it may also significantly influence atmospheric carbon dioxide concentrations.

It is often stated that the thermohaline circulation is the primary reason that the climate of Western Europe is so temperate. An alternate hypothesis claims that this is largely incorrect, and that Europe is warm mostly because it lies downwind of an ocean basin, and because atmospheric waves bring warm air north from the subtropics.

The Antarctic Circumpolar Current encircles that continent, influencing the area's climate and connecting currents in several oceans.

One of the most dramatic forms of weather occurs over the oceans: tropical cyclones (also called "typhoons" and "hurricanes" depending upon where the system forms).

Biology

The ocean has a significant effect on the biosphere. Oceanic evaporation, as a phase of the water cycle, is the source of most rainfall, and ocean temperatures determine climate and wind patterns that affect life on land. Life within the ocean evolved 3 billion years prior to life on land. Both the depth and the distance from shore strongly influence the biodiversity of the plants and animals present in each region.

Lifeforms native to the ocean include:

- Fish;

- Radiata, such as jellyfish (Cnidaria);

- Cetacea, such as whales, dolphins, and porpoises;

- Cephalopods, such as octopus and squid;

- Crustaceans, such as lobsters, shrimp, and krill;

- Marine worms;

- Plankton; and

- Echinoderms, such as brittle stars, starfish, sea cucumbers, and sand dollars.

In addition, many land animals have adapted to living a major part of their life on the oceans. For instance, seabirds are a diverse group of birds that have adapted to a life mainly on the oceans. They feed on marine animals and spend most of their lifetime on water, many only going on land for breeding. Other birds that have adapted to oceans as their living space are penguins, seagulls and pelicans. Seven species of turtles, the sea turtles, also spend most of their time in the oceans.

Gases

Characteristics of Oceanic Gases		
Gas	**Concentration of Seawater, by Mass (in parts per million), for whole Ocean**	**% Dissolved Gas, by Volume, in Seawater at Ocean Surface**
Carbon dioxide (CO_2)	64 to 107	15%
Nitrogen (N_2)	10 to 18	48%
Oxygen (O_2)	0 to 13	36%

Solubility of Oceanic Gases (in terms of mL/L) with Temperature at salinity of 33‰ and atmospheric pressure			
Temperature	**O_2**	**CO_2**	**N_2**
0 °C	8.14	8,700	14.47
10 °C	6.42	8,030	11.59
20 °C	5.26	7,350	9.65
30 °C	4.41	6,600	8.26

Ocean Surface

Generalized characteristics of ocean surface by latitude			
Characteristic	**Oceanic waters in polar regions**	**Oceanic waters in temperate regions**	**Oceanic waters in tropical regions**
Precipitation vs. evaporation	P > E	P > E	E > P
Sea surface temperature in winter	−2 °C	5 to 20 °C	20 to 25 °C
Average salinity	28‰ to 32‰	35‰	35‰ to 37‰
Annual variation of air temperature	≤ 40ªC	10 °C	< 5 °C
Annual variation of water temperature	< 5ªC	10 °C	< 5 °C

Mixing time

The *residence time* is the amount of an element in the ocean divided by the rate at which that element is added to (or removed from) the ocean.

The mean oceanic mixing time is thought to be approximately 1,600 years. If a given element in the ocean stays in the ocean, on average, longer than the oceanic mixing time, then that element is assumed to be homogeneously spread throughout the ocean. As a result, because the major salts have a residence time that is longer than 1,600 years, the ratio of major salts is thought to be unchanging across the ocean. This constant ratio is often referred to as Forchhammer's principle or the principle of constant proportions.

Mean oceanic residence time for various constituents	
Constituent	**Residence time (in years)**
Iron (Fe)	200
Aluminum (Al)	600
Manganese (Mn)	1,300
Water (H_2O)	4,100
Silicon (Si)	20,000
Carbonate (CO_3^{2-})	110,000
Calcium (Ca^{2+})	1,000,000
Sulfate (SO_4^{2-})	11,000,000
Potassium (K^+)	12,000,000
Magnesium (Mg^{2+})	13,000,000
Sodium (Na^+)	68,000,000
Chloride (Cl^-)	100,000,000

Salinity

A zone of rapid salinity increase with depth is called a halocline. The temperature of maximum density of seawater decreases as its salt content increases. Freezing temperature of water decreases with salinity, and boiling temperature of water increases with salinity. Typical seawater freezes at around −1.9 °C at atmospheric pressure. If precipitation exceeds evaporation, as is the case in polar and temperate regions, salinity will be lower. If evaporation exceeds precipitation, as is the case in tropical regions, salinity will be higher. Thus, oceanic waters in polar regions have lower salinity content than oceanic waters in temperate and tropical regions.

Salinity can be calculated using the chlorinity, which is a measure of the total mass of halogen ions (includes fluorine, chlorine, bromine, and iodine) in seawater. By international agreement, the following formula is used to determine salinity:

Salinity (in ‰)=1.80655 x Chlorinity (in ‰)

The average chlorinity is about 19.2‰, and, thus, the average salinity is around 34.7‰

Absorption of Light

Absorption of light in different wavelengths by ocean		
Color: Wavelength (nm)	**Depth wherein 99 percent of wavelength is absorbed (in meters)**	**Percent absorbed in 1 meter of water**
Ultraviolet (UV): 310	31	14.0
Violet (V): 400	107	4.2
Blue (B): 475	254	1.8

Green (G): 525	113	4.0
Yellow (Y): 575	51	8.7
Orange (O): 600	25	16.7
Red (R): 725	4	71.0
Infrared (IR): 800	3	82.0

Economic Value

The oceans are essential to transportation. This is because most of the world's goods move by ship between the world's seaports. Oceans are also the major supply source for the fishing industry. Some of the major harvests are shrimp, fish, crabs, and lobster.

Waves

The motions of the ocean surface, known as undulations or *waves*, are the partial and alternate rising and falling of the ocean surface.

Extraterrestrial Oceans

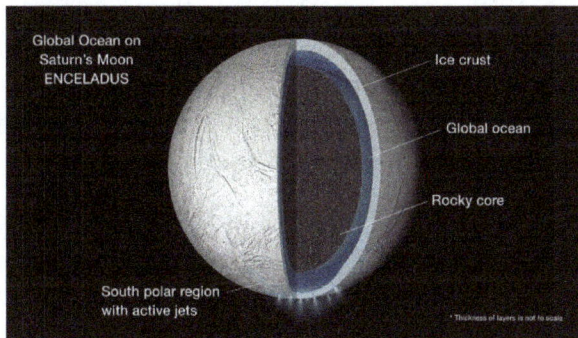

Artist's conception of subsurface ocean of Enceladus confirmed April 3, 2014.

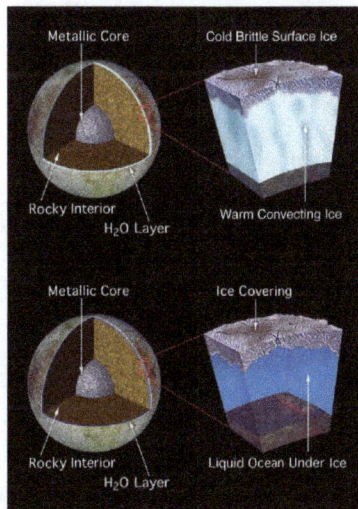

Two models for the composition of Europa predict a large subsurface ocean of liquid water. Similar models have been proposed for other celestial bodies in the Solar System

Although Earth is the only known planet with large stable bodies of liquid water on its surface and the only one in the Solar System, other celestial bodies are thought to have large oceans.

Planets

The gas giants, Jupiter and Saturn, are thought to lack surfaces and instead have a stratum of liquid hydrogen, however their planetary geology is not well understood. The possibility of the ice giants Uranus and Neptune having hot, highly compressed, supercritical water under their thick atmospheres has been hypothesised. Although their composition is still not fully understood, a 2006 study by Wiktorowicz and Ingersall ruled out the possibility of such a water "ocean" existing on Neptune, though some studies have suggested that exotic oceans of liquid diamond are possible.

The Mars ocean hypothesis suggests that nearly a third of the surface of Mars was once covered by water, though the water on Mars is no longer oceanic (much of it residing in the ice caps). The possibility continues to be studied along with reasons for their apparent disappearance. Astronomers think that Venus had liquid water and perhaps oceans in its very early history. If they existed, all later vanished via resurfacing.

Natural Satellites

A global layer of liquid water thick enough to decouple the crust from the mantle is thought to be present on the natural satellites Titan, Europa, Enceladus and, with less certainty, Callisto, Ganymede and Triton. A magma ocean is thought to be present on Io. Geysers have been found on Saturn's moon Enceladus, possibly originating from about 10 kilometers (6.2 mi) deep ocean beneath an ice shell. Other icy moons may also have internal oceans, or may once have had internal oceans that have now frozen.

Large bodies of liquid hydrocarbons are thought to be present on the surface of Titan, although they are not large enough to be considered oceans and are sometimes referred to as *lakes* or seas. The Cassini–Huygens space mission initially discovered only what appeared to be dry lakebeds and empty river channels, suggesting that Titan had lost what surface liquids it might have had. Cassini's more recent fly-by of Titan offers radar images that strongly suggest hydrocarbon lakes exist near the colder polar regions. Titan is thought to have a subsurface liquid-water ocean under the ice and hydrocarbon mix that forms its outer crust.

Dwarf Planets and Trans-Neptunian Objects

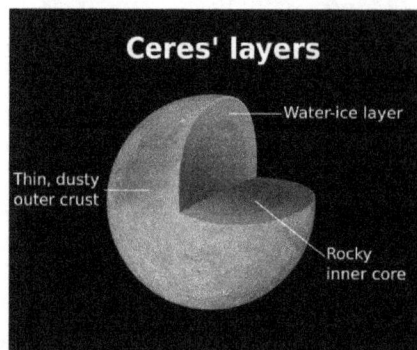

Diagram showing a possible internal structure of Ceres

Ceres appears to be differentiated into a rocky core and icy mantle and may harbour a liquid-water ocean under its surface.

Not enough is known of the larger trans-Neptunian objects to determine whether they are differentiated bodies capable of supporting oceans, although models of radioactive decay suggest that Pluto, Eris, Sedna, and Orcus have oceans beneath solid icy crusts approximately 100 to 180 km thick.

Extrasolar

Rendering of a hypothetical large extrasolar moon with surface liquid-water oceans

Some planets and natural satellites outside the Solar System are likely to have oceans, including possible water ocean planets similar to Earth in the habitable zone or "liquid-water belt". The detection of oceans, even through the spectroscopy method, however is likely extremely difficult and inconclusive.

Theoretical models have been used to predict with high probability that GJ 1214 b, detected by transit, is composed of exotic form of ice VII, making up 75% of its mass, making it an ocean planet.

Other possible candidates are merely speculated based on their mass and position in the habitable zone include planet though little is actually known of their composition. Some scientists speculate Kepler-22b may be an "ocean-like" planet. Models have been proposed for Gliese 581 d that could include surface oceans. Gliese 436 b is speculated to have an ocean of "hot ice". Exomoons orbiting planets, particularly gas giants within their parent star's habitable zone may theoretically have surface oceans.

Terrestrial planets will acquire water during their accretion, some of which will be buried in the magma ocean but most of it will go into a steam atmosphere, and when the atmosphere cools it will collapse on to the surface forming an ocean. There will also be outgassing of water from the mantle as the magma solidifies—this will happen even for planets with a low percentage of their mass composed of water, so "super-Earth exoplanets may be expected to commonly produce water oceans within tens to hundreds of millions of years of their last major accretionary impact."

Non-water Surface Liquids

Oceans, seas, lakes and other bodies of liquids can be composed of liquids other than water, for example the hydrocarbon lakes on Titan. The possibility of seas of nitrogen on Triton was also considered but ruled out. There is evidence that the icy surfaces of the moons Ganymede, Callisto, Europa, Titan and Enceladus are shells floating on oceans of very dense liquid water or water–ammonia. Earth is often called *the* ocean planet because it is 70% covered in water. Extrasolar terrestrial planets that are extremely close to their parent star will be tidally locked and so one half of the planet will be a magma ocean. It is also possible that terrestrial planets had magma oceans at some point during their formation as a result of giant impacts. Hot Neptunes close to their star could lose their atmospheres via hydrodynamic escape, leaving behind their cores with various liquids on the surface. Where there are suitable temperatures and pressures, volatile chemicals that might exist as liquids in abundant quantities on planets include ammonia, argon, carbon disulfide, ethane, hydrazine, hydrogen, hydrogen cyanide, hydrogen sulfide, methane, neon, nitrogen, nitric oxide, phosphine, silane, sulfuric acid, and water.

Supercritical fluids, although not liquids, do share various properties with liquids. Underneath the thick atmospheres of the planets Uranus and Neptune, it is expected that these planets are composed of oceans of hot high-density fluid mixtures of water, ammonia and other volatiles. The gaseous outer layers of Jupiter and Saturn transition smoothly into oceans of supercritical hydrogen. The atmosphere of Venus is 96.5% carbon dioxide, which is a supercritical fluid at its surface.

Arctic Ocean

The Arctic Ocean is the smallest and shallowest of the world's five major oceans. The International Hydrographic Organization (IHO) recognizes it as an ocean, although some oceanographers call it the Arctic Mediterranean Sea or simply the Arctic Sea, classifying it a mediterranean sea or an estuary of the Atlantic Ocean. Alternatively, the Arctic Ocean can be seen as the northernmost part of the all-encompassing World Ocean.

Located mostly in the Arctic north polar region, the Arctic Ocean is almost completely surrounded by Eurasia and North America. It is partly covered by sea ice throughout the year and almost completely in winter. The Arctic Ocean's surface temperature and salinity vary seasonally as the ice cover melts and freezes; its salinity is the lowest on average of the five major oceans, due to low evaporation, heavy fresh water inflow from rivers and streams, and limited connection and outflow to surrounding oceanic waters with higher salinities. The summer shrinking of the ice has been quoted at 50%. The US National Snow and Ice Data Center (NSIDC) uses satellite data to provide a daily record of Arctic sea ice cover and the rate of melting compared to an average period and specific past years.

History

For much of European history, the north polar regions remained largely unexplored and their geography conjectural. Pytheas of Massilia recorded an account of a journey northward in 325 BC, to a land he called "Eschate Thule", where the Sun only set for three hours each day and the water was replaced by a congealed substance "on which one can neither walk nor sail". He was probably describing loose sea ice known today as "growlers" or "bergy bits"; his "Thule" was probably Norway, though the Faroe Islands or Shetland have also been suggested.

Emanuel Bowen's 1780s map of the Arctic features a "Northern Ocean".

Early cartographers were unsure whether to draw the region around the North Pole as land (as in Johannes Ruysch's map of 1507, or Gerardus Mercator's map of 1595) or water (as with Martin Waldseemüller's world map of 1507). The fervent desire of European merchants for a northern passage, the Northern Sea Route or the Northwest Passage, to "Cathay" (China) caused water to win out, and by 1723 mapmakers such as Johann Homann featured an extensive "Oceanus Septentrionalis" at the northern edge of their charts.

The few expeditions to penetrate much beyond the Arctic Circle in this era added only small islands, such as Novaya Zemlya (11th century) and Spitzbergen (1596), though since these were often surrounded by pack-ice, their northern limits were not so clear. The makers of navigational charts, more conservative than some of the more fanciful cartographers, tended to leave the region blank, with only fragments of known coastline sketched in.

This lack of knowledge of what lay north of the shifting barrier of ice gave rise to a number of conjectures. In England and other European nations, the myth of an "Open Polar Sea" was persistent. John Barrow, longtime Second Secretary of the British Admiralty, promoted exploration of the region from 1818 to 1845 in search of this.

In the United States in the 1850s and 1860s, the explorers Elisha Kane and Isaac Israel Hayes both claimed to have seen part of this elusive body of water. Even quite late in the century, the eminent authority Matthew Fontaine Maury included a description of the Open Polar Sea in his textbook *The Physical Geography of the Sea* (1883). Nevertheless, as all the explorers who travelled closer and closer to the pole reported, the polar ice cap is quite thick, and persists year-round.

Fridtjof Nansen was the first to make a nautical crossing of the Arctic Ocean, in 1896. The first surface crossing of the ocean was led by Wally Herbert in 1969, in a dog sled expedition from Alaska to Svalbard, with air support. The first nautical transit of the north pole was made in 1958 by the submarine USS Nautilus, and the first surface nautical transit occurred in 1977 by the icebreaker NS Arktika.

Since 1937, Soviet and Russian manned drifting ice stations have extensively monitored the Arctic Ocean. Scientific settlements were established on the drift ice and carried thousands of kilometres by ice floes.

In World War II, the European region of the Arctic Ocean was heavily contested: the Allied commitment to resupply the Soviet Union via its northern ports was opposed by German naval and air forces.

Geography

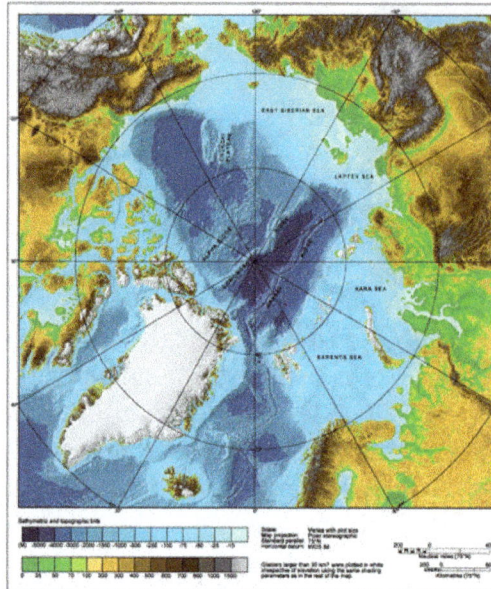

A bathymetric/topographic map of the Arctic Ocean and the surrounding lands.

The Arctic region; of note, the region's southerly border on this map is depicted by a red isotherm, with all territory to the north having an average temperature of less than 10 °C (50 °F) in July.

The Arctic Ocean occupies a roughly circular basin and covers an area of about 14,056,000 km² (5,427,000 sq mi), almost the size of Antarctica. The coastline is 45,390 km (28,200 mi) long. It is surrounded by the land masses of Eurasia, North America, Greenland, and by several islands.

It is generally taken to include Baffin Bay, Barents Sea, Beaufort Sea, Chukchi Sea, East Siberian Sea, Greenland Sea, Hudson Bay, Hudson Strait, Kara Sea, Laptev Sea, White Sea and other tributary bodies of water. It is connected to the Pacific Ocean by the Bering Strait and to the Atlantic Ocean through the Greenland Sea and Labrador Sea.

Countries bordering the Arctic Ocean are: Russia, Norway, Iceland, Greenland, Canada and the United States.

Extent and Major Ports

There are several ports and harbors around the Arctic Ocean

United States

In Alaska, the main ports are Barrow (

71°17′44″N 156°45′59″W71.29556°N 156.76639°W) and Prudhoe Bay (

70°19′32″N 148°42′41″W70.32556°N 148.71139°W).

Canada

In Canada, ships may anchor at Churchill (Port of Churchill) (

58°46′28″N 094°11′37″W58.77444°N 94.19361°W) in Manitoba, Nanisivik (Nanisivik Naval Facility) (

73°04′08″N 084°32′57″W73.06889°N 84.54917°W) in Nunavut, Tuktoyaktuk (

69°26′34″N 133°01′52″W69.44278°N 133.03111°W) or Inuvik (

68°21′42″N 133°43′50″W68.36167°N 133.73056°W) in the Northwest territories.

Greenland

In Greenland, the main port is at Nuuk (Nuuk Port and Harbour) (

64°10′15″N 051°43′15″W64.17083°N 51.72083°W).

Norway

In Norway, Kirkenes (

69°43′37″N 030°02′44″E69.72694°N 30.04556°E) and Vardø (

70°22′14″N 031°06′27″E70.37056°N 31.10750°E) are ports on the mainland. Also, there is Longyearbyen (

78°13′12″N 15°39′00″E78.22000°N 15.65000°E) on the island of Svalbard next to Fram

Strait.

Russia

In Russia, major ports sorted by the different sea areas are:

- Murmansk (68°58′N 033°05′E68.967°N 33.083°E) in the Barents Sea

- Arkhangelsk (64°32′N 040°32′E64.533°N 40.533°E) in the White Sea

- Labytnangi (66°39′26″N 066°25′06″E66.65722°N 66.41833°E) Salekhard (66°32′N 066°36′E66.533°N 66.600°E), Dudinka (69°24′N 086°11′E69.400°N 86.183°E), Igarka (67°28′N 86°35′E67.467°N 86.583°E) and Dikson (73°30′N 080°31′E73.500°N 80.517°E) in the Kara Sea

- Tiksi (71°38′N 128°52′E71.633°N 128.867°E) in the Laptev Sea

- Pevek (69°42′N 170°17′E69.700°N 170.283°E) in the East Siberian Sea

Arctic Shelves

The ocean's Arctic shelf comprises a number of continental shelves, including the Canadian Arctic shelf, underlying the Canadian Arctic Archipelago, and the Russian continental shelf, which is sometimes simply called the "Arctic Shelf" because it is greater in extent. The Russian continental shelf consists of three separate, smaller shelves, the Barents Shelf, Chukchi Sea Shelf and Siberian Shelf. Of these three, the Siberian Shelf is the largest such shelf in the world. The Siberian Shelf holds large oil and gas reserves, and the Chukchi shelf forms the border between Russian and the United States as stated in the USSR–USA Maritime Boundary Agreement. The whole area is subject to international territorial claims.

Underwater Features

An underwater ridge, the Lomonosov Ridge, divides the deep sea North Polar Basin into two oceanic basins: the Eurasian Basin, which is between 4,000 and 4,500 m (13,100 and 14,800 ft) deep, and the Amerasian Basin (sometimes called the North American, or Hyperborean Basin), which is about 4,000 m (13,000 ft) deep. The bathymetry of the ocean bottom is marked by fault block ridges, abyssal plains, ocean deeps, and basins. The average depth of the Arctic Ocean is 1,038 m (3,406 ft). The deepest point is Litke Deep in the Eurasian Basin, at 5,450 m (17,880 ft).

The two major basins are further subdivided by ridges into the Canada Basin (between Alaska/Canada and the Alpha Ridge), Makarov Basin (between the Alpha and Lomonosov Ridges), Nansen Basin (between Lomonosov and Gakkel ridges), and Nansen Basin (Amundsen Basin) (between the Gakkel Ridge and the continental shelf that includes the Franz Josef Land).

Oceanography

Water Flow

In large parts of the Arctic Ocean, the top layer (about 50 m (160 ft)) is of lower salinity and lower temperature than the rest. It remains relatively stable, because the salinity effect on density is bigger than the temperature effect. It is fed by the freshwater input of the big Siberian and Canadian streams (Ob, Yenisei, Lena, Mackenzie), the water of which quasi floats on the saltier, denser, deeper ocean water. Between this lower salinity layer and the bulk of the ocean lies the so-called halocline, in which both salinity and temperature are rising with increasing depth.

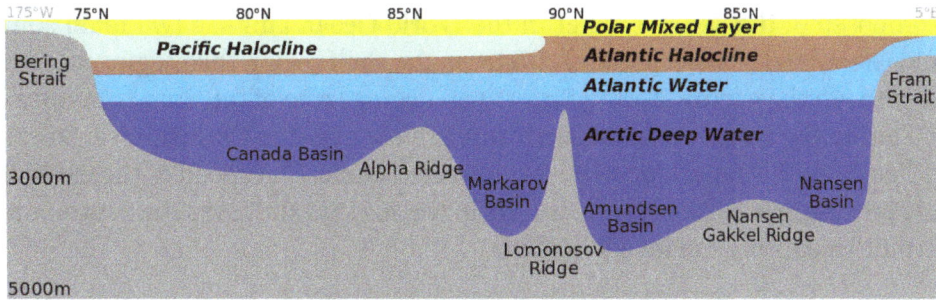

Distribution of the major water mass in the Arctic Ocean. The section sketches the different water masses along a vertical section from Bering Strait over the geographic North Pole to Fram Strait. As the stratification is stable, deeper water masses are more dense than the layers above.

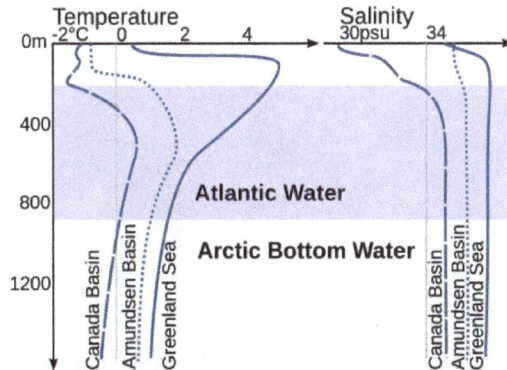

Density structure of the upper 1,200 m (3,900 ft) in the Arctic Ocean. Profiles of temperature and salinity for the Amundsen Basin, the Canadian Basin and the Greenland Sea are sketched in this cartoon.

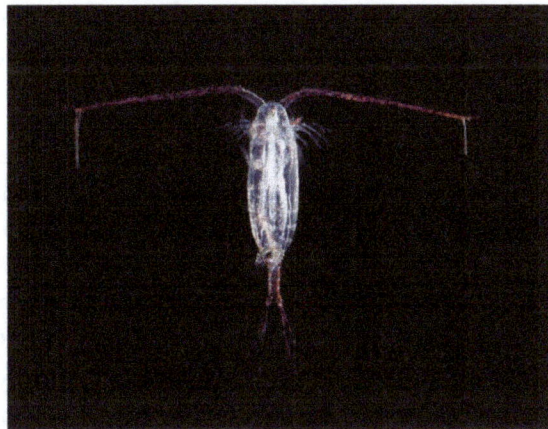

A Copepod

Because of its relative isolation from other oceans, the Arctic Ocean has a uniquely complex system of water flow. It is classified as a mediterranean sea, which as "a part of the world ocean which has only limited communication with the major ocean basins (these being the Pacific, Atlantic, and Indian Oceans) and where the circulation is dominated by thermohaline forcing". The Arctic Ocean has a total volume of 18.07×10^6 km^3, equal to about 1.3% of the World Ocean. Mean surface circulation is predominately cyclonic on the Eurasian side and anticyclonic in the Canadian Basin.

Water enters from both the Pacific and Atlantic Oceans and can be divided into three unique water masses. The deepest water mass is called Arctic Bottom Water and begins around 900 meters depth. It is composed of the densest water in the World Ocean and has two main sources: Arctic shelf water and Greenland Sea Deep Water. Water in the shelf region that begins as inflow from the Pacific passes through the narrow Bering Strait at an average rate of 0.8 Sverdrups and reaches the Chukchi Sea. During the winter, cold Alaskan winds blow over the Chukchi Sea, freezing the surface water and pushing this newly formed ice out to the Pacific. The speed of the ice drift is roughly 1–4 cm/s. This process leaves dense, salty waters in the sea that sink over the continental shelf into the western Arctic Ocean and create a halocline.

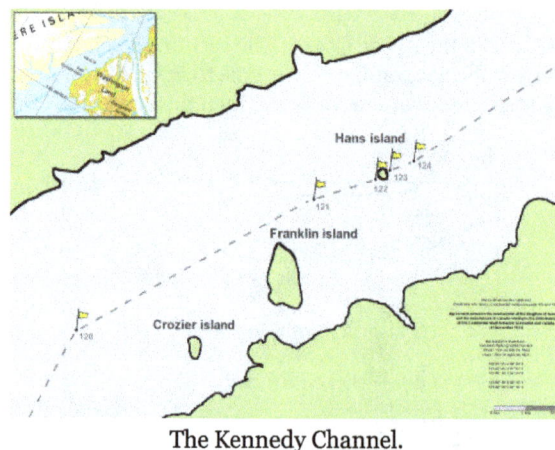

The Kennedy Channel.

This water is met by Greenland Sea Deep Water, which forms during the passage of winter storms. As temperatures cool dramatically in the winter, ice forms and intense vertical convection allows the water to become dense enough to sink below the warm saline water below. Arctic Bottom Water is critically important because of its outflow, which contributes to the formation of Atlantic Deep Water. The overturning of this water plays a key role in global circulation and the moderation of climate.

In the depth range of 150–900 meters is a water mass referred to as Atlantic Water. Inflow from the North Atlantic Current enters through the Fram Strait, cooling and sinking to form the deepest layer of the halocline, where it circles the Arctic Basin counter-clockwise. This is the highest volumetric inflow to the Arctic Ocean, equaling about 10 times that of the Pacific inflow, and it creates the Arctic Ocean Boundary Current. It flows slowly, at about 0.02 m/s. Atlantic Water has the same salinity as Arctic Bottom Water but is much warmer (up to 3 °C). In fact, this water mass is actually warmer than the surface water, and remains submerged only due the role of salinity in density. When water reaches the basin it is pushed by strong winds into a large circular current called the Beaufort Gyre. Water in the Beaufort Gyre is far less saline than that of the Chukchi Sea due to inflow from large Canadian and Siberian rivers.

The final defined water mass in the Arctic Ocean is called Arctic Surface Water and is found from

150–200 meters. The most important feature of this water mass is a section referred to as the sub-surface layer. It is a product of Atlantic water that enters through canyons and is subjected to intense mixing on the Siberian Shelf. As it is entrained, it cools and acts a heat shield for the surface layer. This insulation keeps the warm Atlantic Water from melting the surface ice. Additionally, this water forms the swiftest currents of the Arctic, with speed of around 0.3-0.6 m/s. Complementing the water from the canyons, some Pacific water that does not sink to the shelf region after passing through the Bering Strait also contributes to this water mass.

Waters originating in the Pacific and Atlantic both exit through the Fram Strait between Greenland and Svalbard Island, which is about 2700 meters deep and 350 kilometers wide. This outflow is about 9 Sv. The width of the Fram Strait is what allows for both inflow and outflow on the Atlantic side of the Arctic Ocean. Because of this, it is influenced by the Coriolis force, which concentrates outflow to the East Greenland Current on the western side and inflow to the Norwegian Current on the eastern side. Pacific water also exits along the west coast of Greenland and the Hudson Strait (1-2 Sv), providing nutrients to the Canadian Archipelago.

As noted, the process of ice formation and movement is a key driver in Arctic Ocean circulation and the formation of water masses. With this dependence, the Arctic Ocean experiences variations due to seasonal changes in sea ice cover. Sea ice movement is the result of wind forcing, which is related to a number of meteorological conditions that the Arctic experiences throughout the year. For example, the Beaufort High—an extension of the Siberian High system—is a pressure system that drives the anticyclonic motion of the Beaufort Gyre. During the summer, this area of high pressure is pushed out closer to its Siberian and Canadian sides. In addition, there is a sea level pressure (SLP) ridge over Greenland that drives strong northerly winds through the Fram Strait, facilitating ice export. In the summer, the SLP contrast is smaller, producing weaker winds. A final example of seasonal pressure system movement is the low pressure system that exists over the Nordic and Barents Seas. It is an extension of the Icelandic Low, which creates cyclonic ocean circulation in this area. The low shifts to center over the North Pole in the summer. These variations in the Arctic all contribute to ice drift reaching its weakest point during the summer months. There is also evidence that the drift is associated with the phase of the Arctic Oscillation and Atlantic Multidecadal Oscillation.

Sea Ice

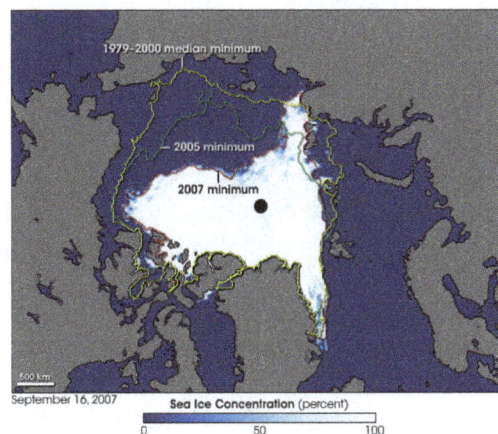

Sea cover in the Arctic Ocean, showing the median, 2005 and 2007 coverage

Much of the Arctic Ocean is covered by sea ice that varies in extent and thickness seasonally. The mean extent of the ice has been decreasing since 1980 from the average winter value of 15,600,000 km² (6,023,200 sq mi) at a rate of 3% per decade. The seasonal variations are about 7,000,000 km² (2,702,700 sq mi) with the maximum in April and minimum in September. The sea ice is affected by wind and ocean currents, which can move and rotate very large areas of ice. Zones of compression also arise, where the ice piles up to form pack ice.

Icebergs occasionally break away from northern Ellesmere Island, and icebergs are formed from glaciers in western Greenland and extreme northeastern Canada. These icebergs pose a hazard to ships, of which the *Titanic* is one of the most famous. Permafrost is found on most islands. The ocean is virtually icelocked from October to June, and the superstructure of ships are subject to icing from October to May. Before the advent of modern icebreakers, ships sailing the Arctic Ocean risked being trapped or crushed by sea ice (although the *Baychimo* drifted through the Arctic Ocean untended for decades despite these hazards).

Climate

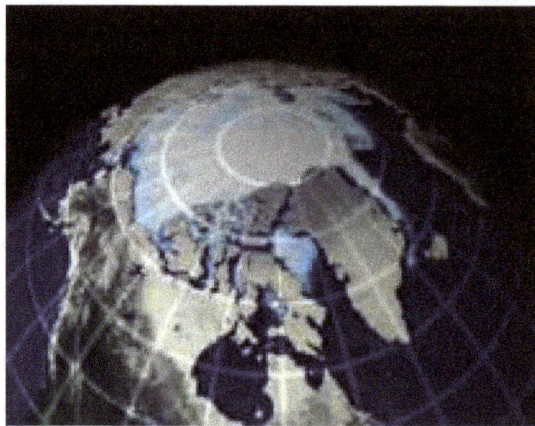

Changes in ice between 1990–1999

Under the influence of the Quaternary glaciation, the Arctic Ocean is contained in a polar climate characterized by persistent cold and relatively narrow annual temperature ranges. Winters are characterized by the polar night, extreme cold, frequent low-level temperature inversions, and stable weather conditions. Cyclones are only common on the Atlantic side. Summers are characterized by continuous daylight (midnight sun), and temperatures can rise above the melting point (0 °C or 32 °F). Cyclones are more frequent in summer and may bring rain or snow. It is cloudy year-round, with mean cloud cover ranging from 60% in winter to over 80% in summer.

The temperature of the surface of the Arctic Ocean is fairly constant, near the freezing point of seawater. Because the Arctic Ocean consists of saltwater, the temperature must reach −1.8 °C (28.8 °F) before freezing occurs.

The density of sea water, in contrast to fresh water, increases as it nears the freezing point and thus it tends to sink. It is generally necessary that the upper 100–150 m (330–490 ft) of ocean water cools to the freezing point for sea ice to form. In the winter the relatively warm ocean water exerts a moderating influence, even when covered by ice. This is one reason why the Arctic does not experience the extreme temperatures seen on the Antarctic continent.

There is considerable seasonal variation in how much pack ice of the Arctic ice pack covers the Arctic Ocean. Much of the Arctic ice pack is also covered in snow for about 10 months of the year. The maximum snow cover is in March or April — about 20 to 50 cm (7.9 to 19.7 in) over the frozen ocean.

The climate of the Arctic region has varied significantly in the past. As recently as 55 million years ago, during the Paleocene–Eocene Thermal Maximum, the region reached an average annual temperature of 10–20 °C (50–68 °F). The surface waters of the northernmost Arctic ocean warmed, seasonally at least, enough to support tropical lifeforms requiring surface temperatures of over 22 °C (72 °F).

Animal and Plant Life

Three polar bears approach USS *Honolulu* near the North Pole.

Endangered marine species in the Arctic Ocean include walruses and whales. The area has a fragile ecosystem which is slow to change and slow to recover from disruptions or damage. Lion's mane jellyfish are abundant in the waters of the Arctic, and the banded gunnel is the only species of gunnel that lives in the ocean.

The Arctic Ocean has relatively little plant life except for phytoplankton. Phytoplankton are a crucial part of the ocean and there are massive amounts of them in the Arctic, where they feed on nutrients from rivers and the currents of the Atlantic and Pacific oceans. During summer, the sun is out day and night, thus enabling the phytoplankton to photosynthesize for long periods of time and reproduce quickly. However, the reverse is true in winter when they struggle to get enough light to survive.

Natural Resources

Petroleum and natural gas fields, placer deposits, polymetallic nodules, sand and gravel aggregates, fish, seals and whales can all be found in abundance in the region.

The political dead zone near the center of the sea is also the focus of a mounting dispute between the United States, Russia, Canada, Norway, and Denmark. It is significant for the global energy market because it may hold 25% or more of the world's undiscovered oil and gas resources.

Environmental Concerns

The Arctic ice pack is thinning, and in many years there is also a seasonal hole in the ozone layer. Reduction of the area of Arctic sea ice reduces the planet's average albedo, possibly resulting in global warming in a positive feedback mechanism. Research shows that the Arctic may become ice free for the first time in human history by 2040.

Warming temperatures in the Arctic may cause large amounts of fresh meltwater to enter the north Atlantic, possibly disrupting global ocean current patterns. Potentially severe changes in the Earth's climate might then ensue.

As the extent of sea ice diminishes and sea level rises, the effect of storms such as the Great Arctic Cyclone of 2012 on open water increases, as does possible salt-water damage to vegetation on shore at locations such as the Mackenzie's river delta as stronger storm surges become more likely.

Other environmental concerns relate to the radioactive contamination of the Arctic Ocean from, for example, Russian radioactive waste dump sites in the Kara Sea and Cold War nuclear test sites such as Novaya Zemlya. In addition, Shell planned to drill exploratory wells in the Chukchi and Beaufort seas during the summer of 2012, which environmental groups filed a lawsuit about in an attempt to protect native communities, endangered wildlife, and the Arctic Ocean in the event of a major oil spill.

On July 16, 2015, five nations (United States of America, Russia, Canada, Norway, Denmark/Greenland) signed a declaration committing to keep their fishing vessels out of a 1.1 million square mile zone in the central Arctic Ocean near the North Pole. The agreement calls for those nations to refrain from fishing there until there is better scientific knowledge about the marine resources and until a regulatory system is in place to protect those resources.

Atlantic Ocean

The Atlantic Ocean is the second largest of the world's oceanic divisions, following the Pacific Ocean. With a total area of about 106,400,000 square kilometres (41,100,000 sq mi), it covers approximately 20 percent of the Earth's surface and about 29 percent of its water surface area. Its name refers to Atlas of Greek mythology, making the Atlantic the "Sea of Atlas".

The oldest known mention of "Atlantic" is in *The Histories* of Herodotus around 450 BC (Hdt. 1.202.4): *Atlantis thalassa*. The term Ethiopic Ocean, derived from Ethiopia, was applied to the southern Atlantic as late as the mid-19th century. Before Europeans discovered other oceans, their term "ocean" was synonymous with the waters beyond the Strait of Gibraltar that are now known as the Atlantic. The early Greeks believed this ocean to be a gigantic river encircling the world.

The Atlantic Ocean occupies an elongated, S-shaped basin extending longitudinally between Eurasia and Africa to the east, and the Americas to the west. As one component of the interconnected global ocean, it is connected in the north to the Arctic Ocean, to the Pacific Ocean in the southwest, the Indian Ocean in the southeast, and the Southern Ocean in the south (other definitions describe the Atlantic as extending southward to Antarctica). The equator subdivides it into the North Atlantic Ocean and South Atlantic Ocean.

Geography

The Atlantic Ocean is bounded on the west by North and South America. It connects to the Arctic Ocean through the Denmark Strait, Greenland Sea, Norwegian Sea and Barents Sea. To the east, the boundaries of the ocean proper are Europe: the Strait of Gibraltar (where it connects with the Mediterranean Sea–one of its marginal seas–and, in turn, the Black Sea, both of which also touch upon Asia) and Africa.

The Atlantic Ocean as seen from the western coast of Portugal

In the southeast, the Atlantic merges into the Indian Ocean. The 20° East meridian, running south from Cape Agulhas to Antarctica defines its border. Some authorities show it extending south to Antarctica, while others show it bounded at the 60° parallel by the Southern Ocean.

In the southwest, the Drake Passage connects it to the Pacific Ocean. The man-made Panama Canal links the Atlantic and Pacific. Besides those mentioned, other large bodies of water that form part of the Atlantic are the Caribbean Sea, the Gulf of Mexico, Hudson Bay, the Mediterranean Sea, the North Sea, the Baltic Sea, and the Celtic Sea.

Covering approximately 22% of Earth's surface, the Atlantic is second in size to the Pacific. With its adjacent seas, it occupies an area of about 106,400,000 square kilometres (41,100,000 sq mi); without them, it has an area of 82,400,000 square kilometres (31,800,000 sq mi). The land that drains into the Atlantic covers four times that of either the Pacific or Indian oceans. The volume of the Atlantic with its adjacent seas is 354,700,000 cubic kilometers (85,100,000 cu mi) and without them 323,600,000 cubic kilometres (77,640,000 cu mi).

The average depth of the Atlantic with its adjacent seas, is 3,339 metres (1,826 fathoms; 10,955 ft); without them it is 3,926 metres (2,147 fathoms; 12,881 ft). The greatest depth, Milwaukee Deep with 8,380 metres (4,580 fathoms; 27,490 ft), is in the Puerto Rico Trench.

Cultural Significance

The Atlantic Ocean was named by the ancient Greeks after either Atlas the Titan or the Atlas Mountains named for him; both involve the concept of holding up the sky. Transatlantic travel played a major role in the expansion of Western civilization into the Americas. It is the Atlantic that separates the "Old World" from the "New World".

In modern times, some idioms refer to the ocean in a humorously diminutive way as the Pond, describing both the geographical and cultural divide between North America and Europe, in particular between the English-speaking nations of both continents. Many Irish or British people refer to the United States and Canada as "across the pond", and vice versa.

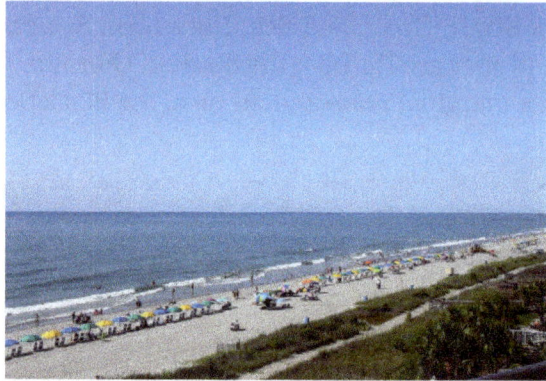

Myrtle Beach, South Carolina view of the Atlantic Ocean.

The "Black Atlantic" refers to the role of this ocean in shaping black people's history, especially through the Atlantic slave trade. Irish migration to the US is meant when the term "The Green Atlantic" is used. The term "Red Atlantic" has been used in reference to the Marxian concept of an Atlantic working class, as well as to the Atlantic experience of indigenous Americans.

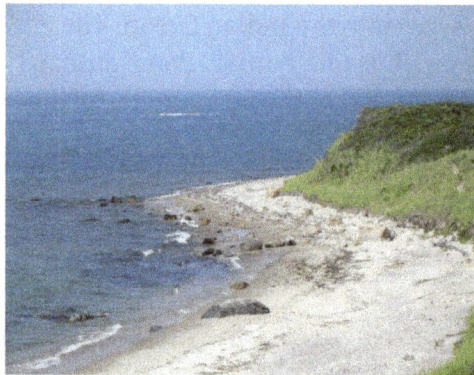

Site of the 1781 shipwreck of the *Culloden* near Montauk, New York.

The Atlantic and Table Mountain in the background, in Cape Town.

Ocean Floor

The principal feature of the bathymetry (bottom topography) is a submarine mountain range called the Mid-Atlantic Ridge. It extends from Iceland in the north to approximately 58° South latitude, reaching a maximum width of about 860 nautical miles (1,590 km; 990 mi). A great rift valley also extends along the ridge over most of its length. The depth of water at the apex of the ridge is less than 2,700 metres (1,500 fathoms; 8,900 ft) in most places, while the bottom of the ridge is three times as deep. Several peaks rise above the water and form islands. The South Atlantic Ocean has an additional submarine ridge, the Walvis Ridge.

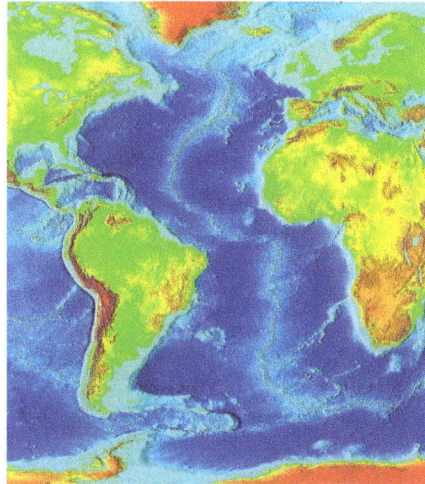

False color map of ocean depth in the Atlantic basin

The Mid-Atlantic Ridge separates the Atlantic Ocean into two large troughs with depths from 3,700–5,500 metres (2,000–3,000 fathoms; 12,100–18,000 ft). Transverse ridges running between the continents and the Mid-Atlantic Ridge divide the ocean floor into numerous basins. Some of the larger basins are the Blake, Guiana, North American, Cape Verde, and Canaries basins in the North Atlantic. The largest South Atlantic basins are the Angola, Cape, Argentina, and Brazil basins.

The deep ocean floor is thought to be fairly flat with occasional deeps, abyssal plains, trenches, seamounts, basins, plateaus, canyons, and some guyots. Various shelves along the margins of the continents constitute about 11% of the bottom topography with few deep channels cut across the continental rise.

Ocean Floor Trenches and Seamounts:

- Puerto Rico Trench, in the North Atlantic, is the deepest trench at 8,605 metres (4,705 fathoms; 28,232 ft)

- Laurentian Abyss is found off the eastern coast of Canada

- South Sandwich Trench reaches a depth of 8,428 metres (4,608 fathoms; 27,651 ft)

- Romanche Trench is located near the equator and reaches a depth of about 7,454 metres (4,076 fathoms; 24,455 ft).

Ocean Sediments are Composed of:

- Terrigenous deposits with land origins, consisting of sand, mud, and rock particles formed by erosion, weathering, and volcanic activity on land washed to sea. These materials are found mostly on the continental shelves and are thickest near large river mouths or off desert coasts.

· Pelagic deposits, which contain the remains of organisms that sink to the ocean floor, include red clays and Globigerina, pteropod, and siliceous oozes. Covering most of the ocean floor and ranging in thickness from 60–3,300 metres (33–1,804 fathoms; 200–10,830 ft) they are thickest in the convergence belts, notably at the Hamilton Ridge and in upwelling zones.

· Authigenic deposits consist of such materials as manganese nodules. They occur where sedimentation proceeds slowly or where currents sort the deposits, such as in the Hewett Curve.

Water Characteristics

Path of the thermohaline circulation. Purple paths represent deep-water currents, while blue paths represent surface currents.

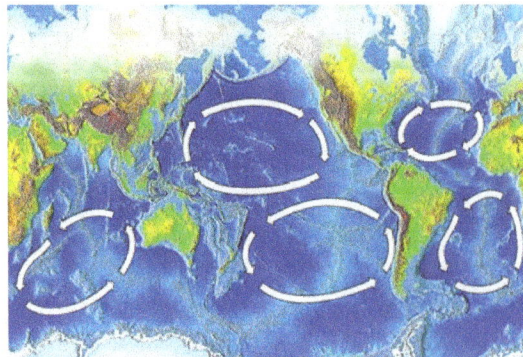

Map of the five major ocean gyres

On average, the Atlantic is the saltiest major ocean; surface water salinity in the open ocean ranges from 33 to 37 parts per thousand (3.3 – 3.7%) by mass and varies with latitude and season. Evaporation, precipitation, river inflow and sea ice melting influence surface salinity values. Although the lowest salinity values are just north of the equator (because of heavy tropical rainfall), in general the lowest values are in the high latitudes and along coasts where large rivers enter. Maximum salinity values occur at about 25° north and south, in subtropical regions with low rainfall and high evaporation.

Surface water temperatures, which vary with latitude, current systems, and season and reflect the latitudinal distribution of solar energy, range from below −2 °C (28 °F) to over 30 °C (86 °F).

Maximum temperatures occur north of the equator, and minimum values are found in the polar regions. In the middle latitudes, the area of maximum temperature variations, values may vary by 7–8 °C (13–14 °F).

The Atlantic Ocean consists of four major water masses. The North and South Atlantic central waters make up the surface. The sub-Antarctic intermediate water extends to depths of 1,000 metres (550 fathoms; 3,300 ft). The North Atlantic Deep Water reaches depths of as much as 4,000 metres (2,200 fathoms; 13,000 ft). The Antarctic Bottom Water occupies ocean basins at depths greater than 4,000 meters.

Within the North Atlantic, ocean currents isolate the Sargasso Sea, a large elongated body of water, with above average salinity. The Sargasso Sea contains large amounts of seaweed and is also the spawning ground for both the European eel and the American eel.

The Coriolis effect circulates North Atlantic water in a clockwise direction, whereas South Atlantic water circulates counter-clockwise. The south tides in the Atlantic Ocean are semi-diurnal; that is, two high tides occur during each 24 lunar hours. In latitudes above 40° North some east-west oscillation occurs.

Climate

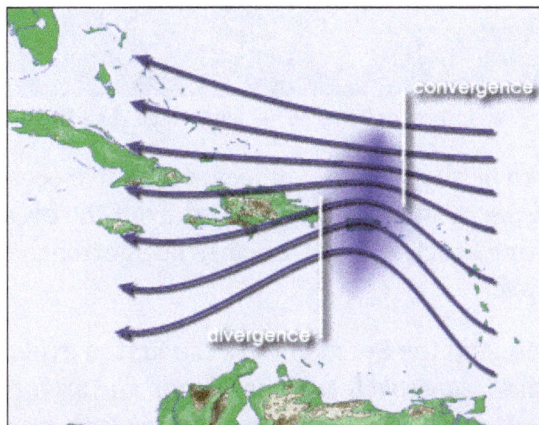

Waves in the trade winds in the Atlantic Ocean—areas of converging winds that move along the same track as the prevailing wind—create instabilities in the atmosphere that may lead to the formation of hurricanes.

Climate is influenced by the temperatures of the surface waters and water currents as well as winds. Because of the ocean's great capacity to store and release heat, maritime climates are more moderate and have less extreme seasonal variations than inland climates. Precipitation can be approximated from coastal weather data and air temperature from water temperatures.

The oceans are the major source of the atmospheric moisture that is obtained through evaporation. Climatic zones vary with latitude; the warmest zones stretch across the Atlantic north of the equator. The coldest zones are in high latitudes, with the coldest regions corresponding to the areas covered by sea ice. Ocean currents influence climate by transporting warm and cold waters to other regions. The winds that are cooled or warmed when blowing over these currents influence adjacent land areas.

The Gulf Stream and its northern extension towards Europe, the North Atlantic Drift, for example, warms the atmosphere of the British Isles and north-western Europe and influences weather and climate as far south as the northern Mediterranean. The cold water currents contribute to heavy fog off the coast of eastern Canada (the Grand Banks of Newfoundland area) and Africa's north-western coast. In general, winds transport moisture and air over land areas. Hurricanes develop in the southern part of the North Atlantic Ocean(Hurricanes are rare in the South Atlantic Ocean). More local particular weather examples could be found in examples such as the Azores High, Benguela Current, and Nor'easter.

History

Animation showing the separation of Pangaea, which formed the Atlantic Ocean known today

The Atlantic Ocean appears to be the second youngest of the five oceans. It did not exist prior to 130 million years ago, when the continents that formed from the breakup of the ancestral super continent Pangaea were drifting apart. The Atlantic has been extensively explored since the earliest settlements along its shores.

The Norsemen, the Portuguese and the Spanish were the first to explore and to cross it systematically, from Europe to the Americas, as well as to its islands and archipelagos, and from the North Atlantic to the South Atlantic. It was after the voyages of Christopher Columbus in 1492, at the service of Castile (later Spain), that the Americas became well known in Europe and European exploration rapidly accelerated, leading to many new trade routes and the colonization of the Americas.

As a result, the Atlantic became and remains the major artery between Europe and the Americas (known as transatlantic trade). Scientific explorations include the Challenger expedition, the German Meteor expedition, Columbia University's Lamont-Doherty Earth Observatory and the United States Navy Hydrographic Office.

Notable Crossings

- Around 600–400 BC, Hanno the Navigator explored West Africa and possibly reached and crossed the Gulf of Guinea and the equator.

- Around 980–982, Norse explorer Erik the Red discovered Greenland, geographically and geologically a part of the Americas.

- Around 1000, Norse explorer Leifur Eríksson, son of Erik the Red, made landfall at Vinland, tentatively identified with the Norse archeological site at L'Anse aux Meadows on the northern tip of Newfoundland on the Atlantic coast of Canada.

- Around 1010, Norse explorers and spouses Þorfinnur *karlsefni* Þórðarson and Guðríður *viðförla* Þorbjarnardóttir led an expedition to Vinland where they begat their son Snorri Þorfinnsson, the first European born in the Americas outside of Greenland.

- In 1419 and 1427, Portuguese navigators reached Madeira and Azores, respectively.

- From 1415 to 1488, Portuguese navigators explored the Western African coast, crossed the Equator, and reached the South Atlantic, the Southern Hemisphere, and the Cape of Good Hope in the southern tip of Africa, entering the Indian Ocean.

- In 1492, Christopher Columbus crossed the Atlantic Ocean and landed on the Bahamas, Cuba and Hispaniola. He made three additional voyages over the next few years, during which he explored the Caribbean coast from Honduras to Venezuela as well as numerous Caribbean islands. These explorations, along with Columbus's attempts to establish a permanent settlement on Hispaniola, led to the European colonization of the Americas and a period of Columbian Exchange that permanently altered human cultures and the environment on both sides of the Atlantic. The establishment of the first transatlantic trade route provided a significant source of revenue to the Crown of Castile, leading to the development of the Spanish Empire. Communicable diseases, unintentionally brought from the Old World to the New World by Europeans, devastated the Amerindian populations, causing the deaths of an estimated 80-95% of the native population of the Americas within 150 years of Columbus's arrival. Columbus also hoped to enslave the native residents of Hispaniola and transport them to Europe; although unsuccessful in this endeavor, his efforts marked the beginning of the transatlantic slave trade that displaced an estimated 11-20 million people from Africa to the Americas over the next several centuries.

- From 1496 to 1498 John Cabot made three voyages to North America from Bristol, landing in Newfoundland and/or possibly the Canadian Maritimes.

- In 1500, Pedro Álvares Cabral reached Brazil.

- In 1519 Ferdinand Magellan sailed from Spain to the South Atlantic, navigating the straits named after him and entering the Pacific Ocean.

- In 1524, Florentine explorer Giovanni da Verrazzano, in the service of the King Francis I of France, discovered the United States of America's east coast.

- In 1534, Jacques Cartier entered the Gulf of St. Lawrence and reached the mouth of the St. Lawrence River.

- In April 1563, Nicolas Barre and 20 other stranded Huguenots were the first to build a (crude) boat in the Americas and sail across the Atlantic. They sailed from Charlesfort, South Carolina to just off the coast of England where they were rescued by an English ship. Though they resorted to cannibalism, seven men survived the voyage, including Barre.

- In 1764, William Harrison (the son of John Harrison) sailed aboard HMS *Tartar*, with the H-4 time piece. The voyage became the basis for the invention of the global system of Longitude.

- In 1858, Cyrus West Field laid the first transatlantic telegraph cable (it quickly failed).

- In 1865, Brunel's ship the SS *Great Eastern* laid the first successful transatlantic telegraph cable.

- In 1870, the small *City of Ragusa* (Dubrovnik) became the first small lifeboat to cross the Atlantic from Cork to Boston with two crew, John Charles Buckley and Nikola Primorac (di Costa).

- In 1896, Frank Samuelsen and George Harbo from Norway became the first people to ever row across the Atlantic Ocean.

- On 15 April 1912 the RMS *Titanic* sank after hitting an iceberg with a loss of more than 1,500 lives.

- On 7 May 1915 the RMS *Lusitania* was torpedoed en route to Queenstown, Ireland, at the loss of 1,198 passengers.

- 1914–1918, during the Atlantic U-boat campaign of World War I, more than 2,100 ships were sunk and 153 U-boats destroyed.

- In 1919, the American NC-4 became the first seaplane to cross the Atlantic (though it made a couple of landings on islands and the sea along the way, and taxied several hundred miles).

- Later in 1919, a British aeroplane piloted by Alcock and Brown made the first non-stop transatlantic flight, from Newfoundland to Ireland.

- In 1921, the British were the first to cross the North Atlantic in an airship.

- In 1922, Portuguese aviators Sacadura Cabral and Gago Coutinho were the First aerial crossing of the South Atlantic on a seaplane connecting Lisbon to Rio de Janeiro.

- In May 1927, Charles Nungesser and François Coli in their aircraft *L'Oiseau Blanc* (*The White Bird*) mysteriously disappeared in an attempt to make the first non-stop transatlantic flight from Paris to New York.

- In 1927, Charles Lindbergh made the first solo non-stop transatlantic flight in an aircraft (between New York City and Paris).

- In 1931, Bert Hinkler made the first solo non-stop transatlantic flight across the South Atlantic in an aircraft.

- In 1932, Amelia Earhart became the first female to make a solo flight across the Atlantic from Harbour Grace, Newfoundland to Derry, Northern Ireland.

- 1939–1945, during World War II, the Battle of the Atlantic resulted in nearly 3,700 ships sunk and 783 U-boats destroyed.

- In 1952, Ann Davison was the first woman to single-handedly sail the Atlantic Ocean.

- In 1965, Robert Manry crossed the Atlantic from the U.S. to England non-stop in a 13.5-foot (4.1-meter) sailboat named "Tinkerbell". Several others also crossed the Atlantic in very small sailboats in the 1960s, none of them non-stop, though.

- In 1969 and 1970 Thor Heyerdahl launched expeditions to cross the Atlantic in boats built from papyrus. He succeeded in crossing the Atlantic from Morocco to Barbados after a two-month voyage of 6,100 km with Ra II in 1970, thus conclusively proving that boats such as the Ra could have sailed with the Canary Current across the Atlantic in prehistoric times.

- In 1980, Gérard d'Aboville was the first man to cross the Atlantic Ocean rowing solo.

- In 1984, Amyr Klink crossed the south atlantic rowing solo from Namibia to Brazil in 100 days.

- In 1984, five Argentines sail in a 10-meter-long raft made from tree trunks named *Atlantis* from Canary Islands and after 52 days 3,000 miles (4,800 km) journey arrived to Venezuela in an attempt to prove travelers from Africa may have crossed the Atlantic before Christopher Columbus.

- In 1994, Guy Delage was the first man to allegedly swim across the Atlantic Ocean (with the help of a kick board, from Cape Verde to Barbados).

- In 1998, Benoît Lecomte was the first man to swim across the northern Atlantic Ocean without a kick board, stopping for only one week in the Azores.

- In 1999, after rowing for 81 days and 4,767 kilometres (2,962 miles), Tori Murden became the first woman to cross the Atlantic Ocean by rowboat alone when she reached Guadeloupe from the Canary Islands.

- In 2003 Alan Priddy and three crew members made a record crossing of the North Atlantic in a RIB from Newfoundland to Scotland, via Greenland and Iceland, in 103 hours.

Ra II, a ship built from papyrus, was successfully sailed across the Atlantic by Thor Heyerdahl proving that it was possible to cross the Atlantic from Africa using such boats in early epochs of history.

Economy

The Atlantic has contributed significantly to the development and economy of surrounding countries. Besides major transatlantic transportation and communication routes, the Atlantic offers abundant petroleum deposits in the sedimentary rocks of the continental shelves. The Atlantic hosts the world's richest fishing resources, especially in the waters covering the shelves. The major fish are cod, haddock, hake, herring, and mackerel.

The most productive areas include the Grand Banks of Newfoundland, the Nova Scotia shelf, Georges Bank off Cape Cod, the Bahama Banks, the waters around Iceland, the Irish Sea, the Dogger Bank of the North Sea, and the Falkland Banks. Eel, lobster, and whales appear in great quantities. Various international treaties attempt to reduce pollution caused by environmental threats such as oil spills, marine debris, and the incineration of toxic wastes at sea.

Terrain

From October to June the surface is usually covered with sea ice in the Labrador Sea, Denmark Strait, and Baltic Sea. A clockwise warm-water gyre occupies the northern Atlantic, and a counter-clockwise warm-water gyre appears in the southern Atlantic. The Mid-Atlantic Ridge, a rugged north-south centerline for the entire Atlantic basin, first discovered by the Challenger Expedition dominates the ocean floor. This was formed by the vulcanism that also formed the ocean floor and the islands rising from it.

The Atlantic has irregular coasts indented by numerous bays, gulfs, and seas. These include the Norwegian Sea, Baltic Sea, North Sea, Labrador Sea, Black Sea, Gulf of Saint Lawrence, Bay of Fundy, Gulf of Maine, Mediterranean Sea, Gulf of Mexico, and Caribbean Sea.

Islands include Newfoundland (including hundreds of surrounding islands), Greenland, Iceland, Faroe Islands, Great Britain, Ireland, Rockall, Sable Island, Azores, St. Pierre and Miquelon, Madeira, Bermuda, Canary Islands, Caribbean Islands (including Greater Antilles, Leeward Islands, Windward Islands, Leeward Antilles), Cape Verde, São Tomé and Príncipe, Annobón Province, Fernando de Noronha, Rocas Atoll, Ascension Island, Saint Helena, Trindade and Martim Vaz, Tristan da Cunha, Gough Island (Also known as Diego Alvarez), Falkland Islands, Tierra del Fuego, South Georgia Island, South Sandwich Islands, and Bouvet Island.

Natural Resources

The Atlantic harbors petroleum and gas fields, fish, marine mammals (seals and whales), sand and gravel aggregates, placer deposits, polymetallic nodules, and precious stones.

Gold deposits are a mile or two under water on the ocean floor, however the deposits are also encased in rock that must be mined through. Currently, there is no cost-effective way to mine or extract gold from the ocean to make a profit.

Natural Hazards

Icebergs are common from February to August in the Davis Strait, Denmark Strait, and the northwestern Atlantic and have been spotted as far south as Bermuda and Madeira. Ships are subject to superstructure icing in the extreme north from October to May. Persistent fog can be a maritime hazard from May to September, as can hurricanes north of the equator (May to December).

Iceberg A22A in the South Atlantic Ocean

The United States' southeast coast has a long history of shipwrecks due to its many shoals and reefs. The Virginia and North Carolina coasts were particularly dangerous.

The Bermuda Triangle is popularly believed to be the site of numerous aviation and shipping incidents because of unexplained and supposedly mysterious causes, but Coast Guard records do not support this belief.

Hurricanes are also a natural hazard in the Atlantic, but mainly in the northern part of the ocean, rarely tropical cyclones form in the southern parts. Hurricanes usually form between 1 June and 30 November of every year.

Current Environmental Issues

Endangered marine species include the manatee, seals, sea lions, turtles, and whales. Drift net fishing can kill dolphins, albatrosses and other seabirds (petrels, auks), hastening the fish stock decline and contributing to international disputes. Municipal pollution comes from the eastern United States, southern Brazil, and eastern Argentina; oil pollution in the Caribbean Sea, Gulf of Mexico, Lake Maracaibo, Mediterranean Sea, and North Sea; and industrial waste and municipal sewage pollution in the Baltic Sea, North Sea, and Mediterranean Sea.

In 2005, there was some concern that warm northern European currents were slowing down.

On 7 June 2006, Florida's wildlife commission voted to take the manatee off the state's endangered species list. Some environmentalists worry that this could erode safeguards for the popular sea creature.

Marine Pollution

Marine pollution is a generic term for the entry into the ocean of potentially hazardous chemicals or particles. The biggest culprits are rivers and with them many agriculture fertilizer chemicals as well as livestock and human waste. The excess of oxygen-depleting chemicals leads to hypoxia and the creation of a dead zone.

Marine debris, which is also known as marine litter, describes human-created waste floating in a body of water. Oceanic debris tends to accumulate at the center of gyres and coastlines, frequently washing aground where it is known as beach litter.

Bordering Countries and Territories

The states (territories in italics) with a coastline on the Atlantic Ocean (excluding the Black, Baltic and Mediterranean Seas) are:

Europe

- *Azores* (PRT)
- Belgium
- Denmark
- *Faroe Islands* (DEN)
- France
- Germany
- *Guernsey* (UK)
- Iceland
- Ireland
- *Isle of Man* (UK)
- *Jersey* (UK)
- Netherlands
- Norway
- Portugal
- Spain
- United Kingdom

Africa

- Angola
- Benin
- *Bouvet Island* (NOR)

- Cameroon
- *Canary Islands* (ESP)
- Cape Verde
- Democratic Republic of the Congo
- Equatorial Guinea
- Gabon
- Gambia
- Ghana
- Guinea
- Guinea-Bissau
- Ivory Coast
- Liberia
- *Madeira* (PRT)
- Mauritania
- Morocco
- Namibia
- Nigeria
- Republic of the Congo
- *Saint Helena, Ascension and Tristan da Cunha* (UK)
- São Tomé and Príncipe
- Senegal
- Sierra Leone
- South Africa
- Togo
- *Western Sahara (claimed by Morocco)* (MAR)

South America

- Argentina
- Brazil

- 🇨🇱 Chile

- 🟨 Colombia

- 🏴 *Falkland Islands* (UK)

- 🟩 *French Guiana* (FRA)

- 🟩 Guyana

- 🏴 *South Georgia and the South Sandwich Islands* (UK)

- 🟩 Suriname

- 🟦 Uruguay

- 🟥 Venezuela

Southern Ocean

The Southern Ocean, also known as the Antarctic Ocean or the Austral Ocean, comprises the southernmost waters of the World Ocean, generally taken to be south of 60° S latitude and encircling Antarctica. As such, it is regarded as the fourth-largest of the five principal oceanic divisions: smaller than the Pacific, Atlantic, and Indian Oceans but larger than the Arctic Ocean. This ocean zone is where cold, northward flowing waters from the Antarctic mix with warmer subantarctic waters.

By way of his voyages in the 1770s, Captain James Cook proved that waters encompassed the southern latitudes of the globe. Since then, geographers have disagreed on the Southern Ocean's northern boundary or even existence, considering the waters part of the Pacific, Atlantic, and Indian Oceans, instead. This remains the current official policy of the International Hydrographic Organization (IHO), since a 2000 revision of its definitions including the Southern Ocean as the waters south of the 60th parallel has not yet been adopted. Others regard the seasonally-fluctuating Antarctic Convergence as the natural boundary.

Definitions and Use

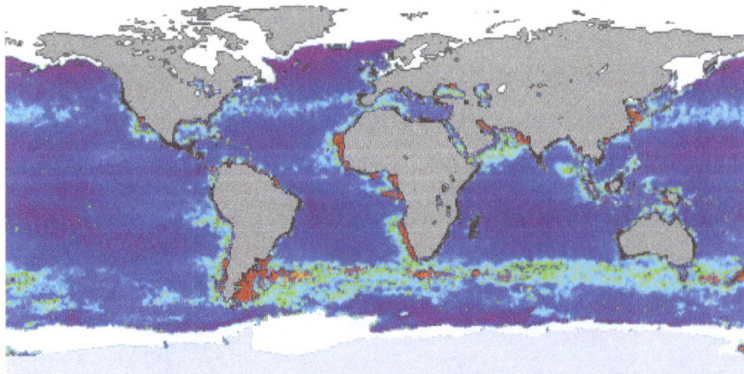

Pulsating pack-ice around Antarctica reflects the various definitions of the Southern Ocean

Borders and names for oceans and seas were internationally agreed when the International Hydrographic Bureau (IHB), the precusor to the IHO, convened the First International Conference on 24 July 1919. The IHO then published these in its *Limits of Oceans and Seas*, the first edition being 1928. Since the first edition, the limits of the Southern Ocean have moved progressively southwards; since 1953, it has been omitted from the official publication and left to local hydrographic offices to determine their own limits. The IHO included the ocean and its definition as the waters south of 60°S in its year 2000 revisions, but this has not been formally adopted, due to continuing impasses over other areas of the text, such as the naming dispute over the Sea of Japan. The 2000 IHO definition, however, was circulated in a draft edition in 2002 and is used by some within the IHO and by some other organizations such as the US Central Intelligence Agency and Merriam-Webster. Australian authorities regard the Southern Ocean as lying immediately south of Australia. The National Geographic Society does not recognize the ocean, depicting it (if at all) in a typeface different from the other world oceans; instead, it shows the Pacific, Atlantic, and Indian Oceans extending to Antarctica on both its print and on line maps. Map publishers using the term Southern Ocean on their maps include Hema Maps and GeoNova.

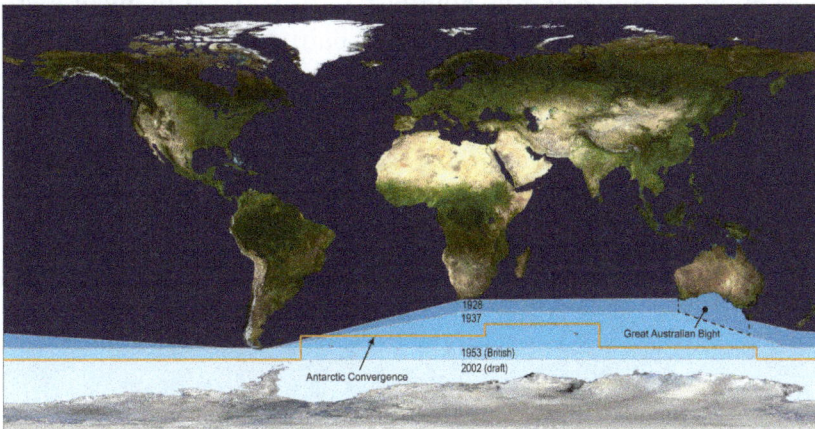

The International Hydrographic Organization's delineation of the "Southern Ocean" has moved steadily southwards since the original 1928 edition of its *Limits of Oceans and Seas*. Australia continues to view the ocean as beginning at its southern coast. The 1953 limits shown are those of Britain, as identified in third edition. Others continue to view the Antarctic Convergence as the natural boundary of the Southern Ocean, regardless of political agreements.

Southern Ocean Antarctica Map.

Pre-20th-century Definitions

"Southern Ocean" is an obsolete name for the Pacific Ocean or South Pacific, coined by Vasco Núñez de Balboa, the first European to sight it, who approached it from the north. The "South Seas" is a less archaic synonym. A 1745 British Act of Parliament established a prize for discovering a Northwest Passage to "the Western and Southern Ocean of *America*".

Authors using "Southern Ocean" to name the waters encircling the unknown southern polar regions used varying limits. James Cook's account of his second voyage implies New Caledonia borders it. Peacock's 1795 *Geographical Dictionary* said it lay "to the southward of America and Africa"; John Payne in 1796 used 40 degrees as the northern limit; the 1827 *Edinburgh Gazeteer* used 50 degrees. The *Family Magazine* in 1835 divided the "Great Southern Ocean" into the "Southern Ocean" and the "Antarctick Ocean" along the Antarctic Circle, with the northern limit of the Southern Ocean being lines joining Cape Horn, the Cape of Good Hope, Van Diemen's Land and the south of New Zealand.

The United Kingdom's South Australia Act 1834 described the waters forming the southern limit of the new colony of South Australia as "the Southern Ocean". The Colony of Victoria's Legislative Council Act of 1881 delimited part of the division of Bairnsdale as "along the New South Wales boundary to the Southern ocean".

1928 Delineation

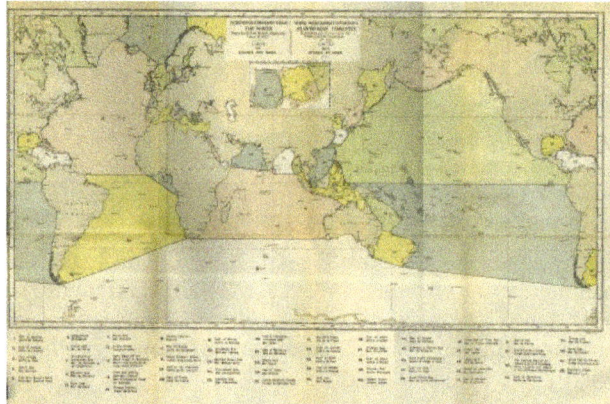

1928 First Edition of *Limits of Oceans and Seas* with original IHO delineation of Southern Ocean abutting land-masses.

In the 1928 first edition of *Limits of Oceans and Seas*, the Southern Ocean was delineated by land-based limits: Antarctica to the south, and South America, Africa, Australia, and Broughton Island, New Zealand to the north.

The detailed land-limits used were from Cape Horn in South America eastwards to Cape Agulhas in Africa, then further eastwards to the southern coast of mainland Australia to Cape Leeuwin, Western Australia. From Cape Leeuwin, the limit then followed eastwards along the coast of mainland Australia to Cape Otway, Victoria, then southwards across Bass Strait to Cape Wickham, King Island, along the west coast of King Island, then the remainder of the way south across Bass Strait to Cape Grim, Tasmania. The limit then followed the west coast of Tasmania southwards to the South East Cape and then went eastwards to Broughton Island, New Zealand, before returning to Cape Horn.

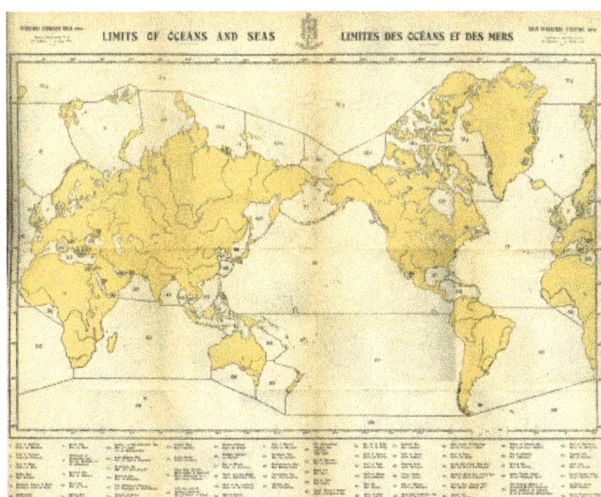

1937 Second Edition of *Limits of Oceans and Seas* showing IHO's pre-1953 delineation of Southern Ocean moved southwards.

1937 Delineation

The northern limits of the Southern Ocean were moved southwards in the IHO's 1937 second edition of the *Limits of Oceans and Seas*. From this edition, much of the ocean's northern limit ceased to abut land masses.

In the second edition, the Southern Ocean then extended from Antarctica northwards to latitude 40°S between Cape Agulhas in Africa (long. 20°E) and Cape Leeuwin in Western Australia (long. 115°E), and extended to latitude 55°S between Auckland Island of New Zealand (165 or 166°E east) and Cape Horn in South America (67°W).

As is discussed in more detail below, prior to the 2002 (draft) edition the limits of oceans explicitly excluded the seas lying within each of them. The Great Australian Bight was unnamed in the 1928 edition, and delineated as shown in the figure above in the 1937 edition. It therefore encompassed former Southern Ocean waters (as designated in 1928) but was technically not inside any of the three adjacent oceans by 1937. In the 2002 draft edition, the IHO have designated 'seas' as being subdivisions within 'oceans', so the Bight would have still been within the Southern Ocean in 1937 if the 2002 convention were in place then. To perform direct comparisons of current and former limits of oceans (for example to compare surface areas) it is necessary to consider, or at least be aware of, how the 2002 change in IHO terminology for 'seas' can affect the comparison.

1953 Delineation

The Southern Ocean did not appear in the 1953 third edition and a note in the publication read:

The Antarctic or Southern Ocean has been omitted from this publication as the majority of opinions received since the issue of the 2nd Edition in 1937 are to the effect that there exists no real justification for applying the term Ocean to this body of water, the northern limits of which are difficult to lay down owing to their seasonal change. The limits of the Atlantic, Pacific and Indian Oceans have therefore been extended South to the Antarctic Continent.

Hydrographic Offices who issue separate publications dealing with this area are therefore left to decide their own northern limits (Great Britain uses Latitude of 55 South.)

Instead, in the IHO 1953 publication, the Atlantic, Indian and Pacific Oceans were extended southward, the Indian and Pacific Oceans (which had not previously touched pre 1953, as per the first and second editions) now abutted at the meridian of South East Cape, and the southern limits of the Great Australian Bight and the Tasman Sea were moved northwards.

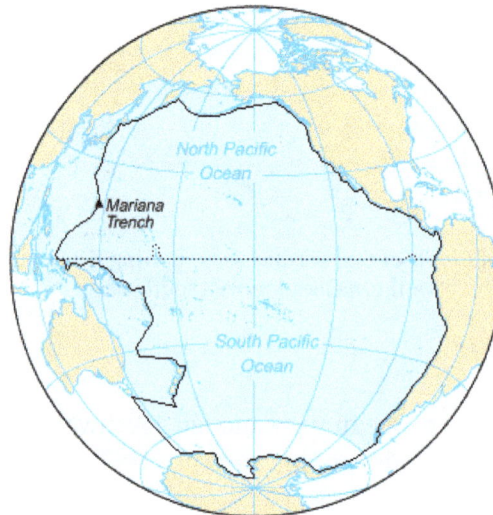

The Pacific Ocean as example of terminology concerning seas: the area inside the black line includes the seas included in the Pacific Ocean prior to 2002 and the darker blue areas are its informal current borders, following the recreation of the Southern Ocean and the reinclusion of marginal seas.

2002 (Draft) Delineation

The IHO readdressed the question of the Southern Ocean in a survey in 2000. Of its 68 member nations, 28 responded, and all responding members except Argentina agreed to redefine the ocean, reflecting the importance placed by oceanographers on ocean currents. The proposal for the name *Southern Ocean* won 18 votes, beating the alternative *Antarctic Ocean*. Half of the votes supported a definition of the ocean's northern limit at 60°S (with no land interruptions at this latitude), with the other 14 votes cast for other definitions, mostly 50°S, but a few for as far north as 35°S.

A draft fourth edition of *Limits of Oceans and Seas* was circulated to IHO member states in August 2002 (sometimes referred to as the "2000 edition" as it summarized the progress to 2000). It has yet to be published due to 'areas of concern' by several countries relating to various naming issues around the world – primarily the Sea of Japan naming dispute – and there have been various changes, 60 seas were given new names, and even the name of the publication was changed. A reservation had also been lodged by Australia regarding the Southern Ocean limits. Effectively, the 3rd edition (which did not delineate the Southern Ocean leaving delineation to local hydrographic offices) has yet to be superseded.

Despite this, the 4th edition definition has partial *de facto* usage by many nations, scientists and organisations such as the U.S. (the Central Intelligence Agency World Factbook uses "Southern Ocean" but none of the other new sea names within the "Southern Ocean" such as "Cosmonauts Sea") and Merriam-Webster, scientists and nations – and even by some within the IHO. Some nations' hydrographic offices have defined their own boundaries; the United Kingdom used the 55°S parallel for example. Other organisations favour more northerly limits for the Southern Ocean. For example, the *Encyclopædia Britannica* describes the Southern Ocean as extending as far north as South America, and confers great significance on the Antarctic Convergence, yet its description of the Indian Ocean contradicts this, describing the Indian Ocean as extending south to Antarctica.

Other sources, such as the National Geographic Society, show the Atlantic, Pacific and Indian Oceans as extending to Antarctica on its maps, although articles on the National Geographic web site have begun to reference the Southern Ocean.

In Australia, cartographical authorities defined the Southern Ocean as including the entire body of water between Antarctica and the south coasts of Australia and New Zealand. This delineation is basically the same as the original (first) edition of the IHO publication and effectively the same as the second edition. In the second edition, the Great Australian Bight was defined as the only geographical entity between the Australian coast and the Southern Ocean. Coastal maps of Tasmania and South Australia label the sea areas as *Southern Ocean*, while Cape Leeuwin in Western Australia is described as the point where the Indian and Southern Oceans meet.

A radical shift from past IHO practices (1928–1953) was also seen in the 2002 draft edition when the IHO delineated 'seas' as being subdivisions that lay within the boundaries of 'oceans'. While the IHO are often considered the authority for such conventions, the shift brought them into line with the practices of other publications (e.g. the CIA *World Fact Book*) which already adopted the principle that seas are contained within oceans. This difference in practice is markedly seen for the Pacific Ocean in the adjacent figure. Thus, for example, previously the Tasman Sea between Australia and New Zealand was not regarded by the IHO as being part of the Pacific, but as of the 2002 draft edition it is.

The new delineation of seas being subdivisions of oceans has avoided the need to interrupt the northern boundary of the Southern Ocean where intersected by Drake Passage which includes all of the waters from South America to the Antarctic coast, nor interrupt it for the Scotia Sea, which also extends below the 60th parallel south. The new delineation of seas has also meant that the long-time named seas around Antarctica, excluded from the 1953 edition (the 1953 map did not even extend that far south), are 'automatically' part of the Southern Ocean.

History of Exploration

The Unknown Southern Land

Exploration of the Southern Ocean was inspired by a belief in the existence of a *Terra Australis*—a vast continent in the far south of the globe to "balance" the northern lands of Eurasia and North Africa—which had existed since the times of Ptolemy. The doubling of the Cape of Good Hope in 1487 by Bartolomeu Dias first brought explorers within touch of the Antarctic cold, and proved that there was an ocean separating Africa from any Antarctic land that might exist. Ferdinand Magellan, who passed through the Strait of Magellan in 1520, assumed that the islands of Tierra del Fuego to the south were an extension of this unknown southern land. In 1564, Abraham Ortelius published his first map, *Typus Orbis Terrarum*, an eight-leaved wall map of the world, on which he identified the *Regio Patalis* with *Locach* as a northward extension of the *Terra Australis*, reaching as far as New Guinea.

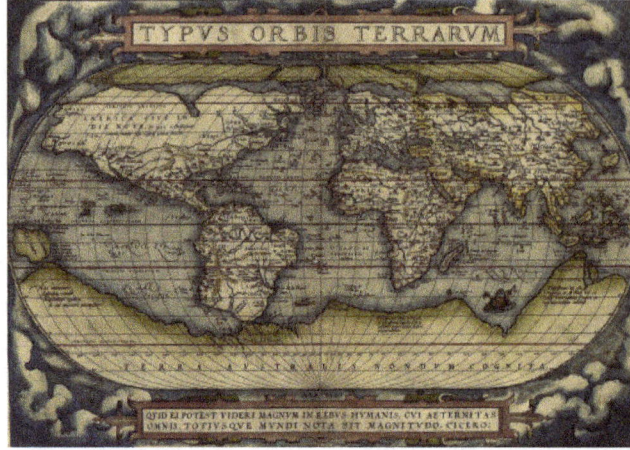

1564 *Typus Orbis Terrarum*, a map by Abraham Ortelius showed the imagined link between the proposed continent of Antarctica and South America.

European geographers continued to connect the coast of Tierra del Fuego with the coast of New Guinea on their globes, and allowing their imaginations to run riot in the vast unknown spaces of the south Atlantic, south Indian and Pacific oceans they sketched the outlines of the *Terra Australis Incognita* ("Unknown Southern Land"), a vast continent stretching in parts into the tropics. The search for this great south land was a leading motive of explorers in the 16th and the early part of the 17th centuries.

The Spaniard Gabriel de Castilla, who claimed having sighted "snow-covered mountains" beyond the 64° S in 1603, is recognized as the first explorer that discovered the continent of Antarctica, although he was ignored in his time.

Portrait of Edmund Halley by Thomas Murray, c. 1687

In 1606, Pedro Fernández de Quirós took possession for the king of Spain all of the lands he had discovered in Australia del Espiritu Santo (the New Hebrides) and those he would discover "even to the Pole".

Francis Drake, like Spanish explorers before him, had speculated that there might be an open channel south of Tierra del Fuego. When Willem Schouten and Jacob Le Maire discovered the southern extremity of Tierra del Fuego and named it Cape Horn in 1615, they proved that the Tierra del Fuego archipelago was of small extent and not connected to the southern land, as previously thought. Subsequently, in 1642, Abel Tasman showed that even New Holland (Australia) was separated by sea from any continuous southern continent.

South of the Antarctic Convergence

The visit to South Georgia by Anthony de la Roché in 1675 was the first ever discovery of land south of the Antarctic Convergence i.e. in the Southern Ocean / Antarctic. Soon after the voyage cartographers started to depict 'Roché Island', honouring the discoverer. James Cook was aware of la Roché's discovery when surveying and mapping the island in 1775.

Edmond Halley's voyage in HMS *Paramour* for magnetic investigations in the South Atlantic met the pack ice in 52° S in January 1700, but that latitude (he reached 140 mi off the north coast of South Georgia) was his farthest south. A determined effort on the part of the French naval officer Jean-Baptiste Charles Bouvet de Lozier to discover the "South Land" – described by a half legendary "sieur de Gonneyville" – resulted in the discovery of Bouvet Island in 54°10′ S, and in the navigation of 48° of longitude of ice-cumbered sea nearly in 55° S in 1730 .

In 1771, Yves Joseph Kerguelen sailed from France with instructions to proceed south from Mauritius in search of "a very large continent." He lighted upon a land in 50° S which he called South France, and believed to be the central mass of the southern continent. He was sent out again to complete the exploration of the new land, and found it to be only an inhospitable island which he renamed the Isle of Desolation, but which was ultimately named after him.

South of the Antarctic Circle

Famous official portrait of Captain James Cook who proved that waters encompassed the southern latitudes of the globe. *"He holds his own chart of the Southern Ocean on the table and his right hand points to the east coast of Australia on it."*

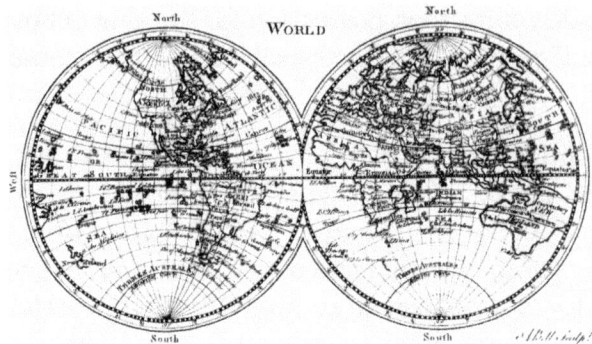

Map from 1771, showing "Terres Australes" (sic) label without any charted landmass.

Painting of James Weddell's second expedition in 1823, depicting the brig *Jane* and the cutter *Beaufroy*.

The obsession of the undiscovered continent culminated in the brain of Alexander Dalrymple, the brilliant and erratic hydrographer who was nominated by the Royal Society to command the Transit of Venus expedition to Tahiti in 1769. The command of the expedition was given by the admiralty to Captain James Cook. Sailing in 1772 with the *Resolution*, a vessel of 462 tons under his own command and the *Adventure* of 336 tons under Captain Tobias Furneaux, Cook first searched in vain for Bouvet Island, then sailed for 20 degrees of longitude to the westward in latitude 58° S, and then 30° eastward for the most part south of 60° S, a lower southern latitude than had ever been voluntarily entered before by any vessel. On 17 January 1773 the Antarctic Circle was crossed for the first time in history and the two ships reached 67° 15' S by 39° 35' E, where their course was stopped by ice.

Cook then turned northward to look for French Southern and Antarctic Lands, of the discovery of which he had received news at Cape Town, but from the rough determination of his longitude by Kerguelen, Cook reached the assigned latitude 10° too far east and did not see it. He turned south again and was stopped by ice in 61° 52' S by 95° E and continued eastward nearly on the parallel of 60° S to 147° E. On 16 March, the approaching winter drove him northward for rest to New Zealand and the tropical islands of the Pacific. In November 1773, Cook left New Zealand, having parted company with the *Adventure*, and reached 60° S by 177° W, whence he sailed eastward keeping as far south as the floating ice allowed. The Antarctic Circle was crossed on 20 December and Cook remained south of it for three days, being compelled after reaching 67° 31' S to stand north again in 135° W.

A long detour to 47° 50' S served to show that there was no land connection between New Zealand

and Tierra del Fuego. Turning south again, Cook crossed the Antarctic Circle for the third time at 109° 30′ W before his progress was once again blocked by ice four days later at 71° 10′ S by 106° 54′ W. This point, reached on 30 January 1774, was the farthest south attained in the 18th century. With a great detour to the east, almost to the coast of South America, the expedition regained Tahiti for refreshment. In November 1774, Cook started from New Zealand and crossed the South Pacific without sighting land between 53° and 57° S to Tierra del Fuego; then, passing Cape Horn on 29 December, he rediscovered Roché Island renaming it Isle of Georgia, and discovered the South Sandwich Islands (named *Sandwich Land* by him), the only ice-clad land he had seen, before crossing the South Atlantic to the Cape of Good Hope between 55° and 60°. He thereby laid open the way for future Antarctic exploration by exploding the myth of a habitable southern continent. Cook's most southerly discovery of land lay on the temperate side of the 60th parallel, and he convinced himself that if land lay farther south it was practically inaccessible and without economic value.

Voyagers rounding the Horn frequently met with contrary winds and were driven southward into snowy skies and ice-encumbered seas; but so far as can be ascertained none of them before 1770 reached the Antarctic Circle, or knew it, if they did.

In a voyage from 1822 to 1824, James Weddell commanded the 160-ton brig *Jane*, accompanied by his second ship *Beaufoy* captained by Matthew Brisbane. Together they sailed to the South Orkneys where sealing proved disappointing. They turned south in the hope of finding a better sealing ground. The season was unusually mild and tranquil, and on 20 February 1823 the two ships reached latitude 74°15′ S and longitude 34°16′45″ W the southernmost position any ship had ever reached up to that time. A few icebergs were sighted but there was still no sight of land, leading Weddell to theorize that the sea continued as far as the South Pole. Another two days' sailing would have brought him to Coat's Land (to the east of the Weddell Sea) but Weddell decided to turn back.

First Sighting of Land

Admiral von Bellingshausen

The first land south of the parallel 60° south latitude was discovered by the Englishman William Smith, who sighted Livingston Island on 19 February 1819. A few months later Smith returned to explore the other islands of the South Shetlands archipelago, landed on King George Island, and claimed the new territories for Britain.

In the meantime, the Spanish Navy ship *San Telmo* sank in September 1819 when trying to cross Cape Horn. Parts of her wreckage were found months later by sealers on the north coast of Livingston Island (South Shetlands). It is unknown if some survivor managed to be the first to set foot on these Antarctic islands.

The first confirmed sighting of mainland Antarctica cannot be accurately attributed to one single person. It can, however, be narrowed down to three individuals. According to various sources, three men all sighted the ice shelf or the continent within days or months of each other: von Bellingshausen, a captain in the Russian Imperial Navy; Edward Bransfield, a captain in the British navy; and Nathaniel Palmer, an American sealer out of Stonington, Connecticut. It is certain that the expedition, led by von Bellingshausen and Lazarev on the ships *Vostok* and *Mirny,* reached a point within 32 km (20 mi) from Princess Martha Coast and recorded the sight of an ice shelf at

69°21′28″S 2°14′50″W69.35778°S 2.24722°W that became known as the Fimbul ice shelf. On 30 January 1820, Bransfield sighted Trinity Peninsula, the northernmost point of the Antarctic mainland, while Palmer sighted the mainland in the area south of Trinity Peninsula in November 1820. Von Bellingshausen's expedition also discovered Peter I Island and Alexander I Island, the first islands to be discovered south of the circle.

Antarctic Expeditions

USS Vincennes at Disappointment Bay, Antarctica in Early 1840.

1911 South Polar Regions exploration map

In December 1839, as part of the United States Exploring Expedition of 1838–42 conducted by the United States Navy (sometimes called "the Wilkes Expedition"), an expedition sailed from Sydney, Australia, on the sloops-of-war USS *Vincennes* and USS *Peacock* , the brig USS *Porpoise* , the full-rigged ship *Relief*, and two schooners *Sea Gull* and USS *Flying Fish* . They sailed into the Antarctic Ocean, as it was then known, and reported the discovery "of an Antarctic continent west of the Balleny Islands" on 25 January 1840. That part of Antarctica was later named "Wilkes Land", a name it maintains to this day.

Explorer James Clark Ross passed through what is now known as the Ross Sea and discovered Ross Island (both of which were named for him) in 1841. He sailed along a huge wall of ice that was later named the Ross Ice Shelf. Mount Erebus and Mount Terror are named after two ships from his expedition: HMS *Erebus* and *Terror*.

Frank Hurley, *As time wore on it became more and more evident that the ship was doomed* (The Endurance trapped in pack ice), National Library of Australia.

The Imperial Trans-Antarctic Expedition of 1914, led by Ernest Shackleton, set out to cross the continent via the pole, but their ship, the *Endurance*, was trapped and crushed by pack ice before they even landed. The expedition members survived after an epic journey on sledges over pack ice to Elephant Island. Then Shackleton and five others crossed the Southern Ocean, in an open boat called *James Caird*, and then trekked over South Georgia to raise the alarm at the whaling station Grytviken.

In 1946, US Navy Rear Admiral Richard Evelyn Byrd and more than 4,700 military personnel visited the Antarctic in an expedition called Operation *Highjump*. Reported to the public as a scientific mission, the details were kept secret and it may have actually been a training or testing mission for the military. The expedition was, in both military or scientific planning terms, put together very quickly. The group contained an unusually high amount of military equipment, including an aircraft carrier, submarines, military support ships, assault troops and military vehicles. The expedition was planned to last for eight months but was unexpectedly terminated after only two months. With the exception of some eccentric entries in Admiral Byrd's diaries, no real explanation for the early termination has ever been officially given.

Captain Finn Ronne, Byrd's executive officer, returned to Antarctica with his own expedition in 1947–1948, with Navy support, three planes, and dogs. Ronne disproved the notion that the continent was divided in two and established that East and West Antarctica was one single continent, i.e. that the Weddell Sea and the Ross Sea are not connected. The expedition explored and mapped large parts of Palmer Land and the Weddell Sea coastline, and identified the Ronne Ice Shelf, named by Ronne after his wife Edith "Jackie" Ronne. Ronne covered 3,600 miles by ski and dog sled—more than any other explorer in history. The Ronne Antarctic Research Expedition discovered and mapped the last unknown coastline in the world and was the first Antarctic expedition to ever include women.

Recent History

MS *Explorer* in Antarctica in January 1999. She sank on 23 November 2007 after hitting an iceberg.

The Antarctic Treaty was signed on 1 December 1959 and came into force on 23 June 1961. Among other provisions, this treaty limits military activity in the Antarctic to the support of scientific research.

The first person to sail single-handed to Antarctica was the New Zealander David Henry Lewis, in 1972, in a 10-metre steel sloop *Ice Bird*.

A baby, named Emilio Marcos de Palma, was born near Hope Bay on 7 January 1978, becoming the first baby born on the continent. He also was born further south than anyone in history.

The MS *Explorer* was a cruise ship operated by the Swedish explorer Lars-Eric Lindblad. Observers point to the *Explorer's* 1969 expeditionary cruise to Antarctica as the frontrunner for today's sea-based tourism in that region. The *Explorer* was the first cruise ship used specifically to sail the icy waters of the Antarctic Ocean and the first to sink there when she struck an unidentified submerged object on 23 November 2007, reported to be ice, which caused a 10 by 4 inches (25 by 10 cm) gash in the hull. The *Explorer* was abandoned in the early hours of 23 November 2007 after taking on water near the South Shetland Islands in the Southern Ocean, an area which is usually stormy but was calm at the time. The *Explorer* was confirmed by the Chilean Navy to have sunk at approximately position: 62° 24′ South, 57° 16′ West, in roughly 600 m of water.

Geography

The Southern Ocean, geologically the youngest of the oceans, was formed when Antarctica and South America moved apart, opening the Drake Passage, roughly 30 million years ago. The separation of the continents allowed the formation of the Antarctic Circumpolar Current.

With a northern limit at 60°S, the Southern Ocean differs from the other oceans in that its largest boundary, the northern boundary, does not abut a landmass (as it did with the first edition of *Limits of Oceans and Seas*). Instead, the northern limit is with the Atlantic, Indian and Pacific Oceans.

One reason for considering it as a separate ocean stems from the fact that much of the water of the Southern Ocean differs from the water in the other oceans. Water gets transported around the Southern Ocean fairly rapidly because of the Antarctic Circumpolar Current which circulates around Antarctica. Water in the Southern Ocean south of, for example, New Zealand, resembles the water in the Southern Ocean south of South America more closely than it resembles the water in the Pacific Ocean.

The Southern Ocean has typical depths of between 4,000 and 5,000 m (13,000 and 16,000 ft) over most of its extent with only limited areas of shallow water. The Southern Ocean's greatest depth of 7,236 m (23,740 ft) occurs at the southern end of the South Sandwich Trench, at 60°00'S, 024°W. The Antarctic continental shelf appears generally narrow and unusually deep, its edge lying at depths up to 800 m (2,600 ft), compared to a global mean of 133 m (436 ft).

Equinox to equinox in line with the sun's seasonal influence, the Antarctic ice pack fluctuates from an average minimum of 2.6 million square kilometres (1.0×10^6 sq mi) in March to about 18.8 million square kilometres (7.3×10^6 sq mi) in September, more than a sevenfold increase in area.

Sub-divisions of the Southern Ocean

An iceberg being pushed out of a shipping lane USS *Burton Island* (AGB-1), USS *Atka* (AGB-3), and USS *Glacier* (AGB-4) pushing an iceberg out of a channel in the "Silent Land" near McMurdo Station, Antarctica, 1965

Sub-divisions of oceans are geographical features such as "seas", "straits", "bays", "channels", and "gulfs". There are many sub-divisions of the Southern Ocean defined in the never-approved 2002 draft fourth edition of the IHO publication *Limits of Oceans and Seas*. In clockwise order these include (with IHO sub-division chartlet numbers in parenthesis) the Weddell Sea (10.1), the Lazarev Sea (10.2), the Riiser-Larsen Sea (10.3), the Cosmonauts Sea (10.4), the Cooperation Sea (10.5), the Davis Sea (10.6), Tryoshinikova Gulf (10.6.1), the Mawson Sea (10.7), the Dumont D'Urville Sea (10.8), the Somov Sea (10.9), the Ross Sea (10.10), McMurdo Sound (10.10.1), the Amundsen Sea (10.11), the Bellingshausen Sea (10.12), part of the Drake Passage (10.13), Bransfield Strait (10.14) and part of the Scotia Sea (4.2). A number of these such as the 2002 Russian-proposed "Consmonauts Sea", "Cooperation Sea", and "Somov (mid-1950s Russian polar explorer) Sea" are not included in the 1953 IHO document which remains currently in force, because they received their names largely originated from 1962 onward. Leading geographic authorities and atlases do not use these latter three names, including the 2014 10th edition World Atlas from the United States' National Geographic Society and the 2014 12th edition of the British Times Atlas of the World, but Soviet and Russian-issued maps do.

Natural Resources

Manganese nodule

The Southern Ocean probably contains large, and possibly giant, oil and gas fields on the continental margin. Placer deposits, accumulation of valuable minerals such as gold, formed by gravity separation during sedimentary processes are also expected to exist in the Southern Ocean.

Manganese nodules are expected to exist in the Southern Ocean. Manganese nodules are rock concretions on the sea bottom formed of concentric layers of iron and manganese hydroxides around a core. The core may be microscopically small and is sometimes completely transformed into manganese minerals by crystallization. Interest in the potential exploitation of polymetallic nodules generated a great deal of activity among prospective mining consortia in the 1960s and 1970s.

The icebergs that form each year around in the Southern Ocean hold enough fresh water to meet the needs of every person on Earth for several months. For several decades there have been proposals, none yet to be feasible or successful, to tow Southern Ocean icebergs to more arid northern regions (such as Australia) where they can be harvested.

Natural Hazards

Icebergs can occur at any time of year throughout the ocean. Some may have drafts up to several hundred meters; smaller icebergs, iceberg fragments and sea-ice (generally 0.5 to 1 m thick) also pose problems for ships. The deep continental shelf has a floor of glacial deposits varying widely over short distances.

Sailors know latitudes from 40 to 70 degrees south as the "Roaring Forties", "Furious Fifties" and "Shrieking Sixties" due to high winds and large waves that form as winds blow around the entire globe unimpeded by any land-mass. Icebergs, especially in May to October, make the area even more dangerous. The remoteness of the region makes sources of search and rescue scarce.

Physical Oceanography

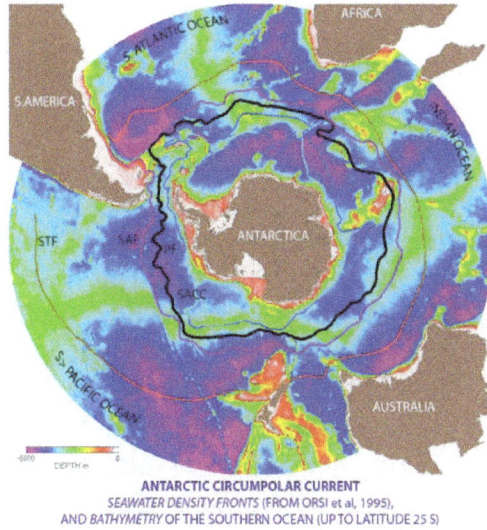

ANTARCTIC CIRCUMPOLAR CURRENT
SEAWATER DENSITY FRONTS (FROM ORSI et al, 1995),
AND *BATHYMETRY* OF THE SOUTHERN OCEAN (UP TO LATITUDE 25 S)

The Antarctic Circumpolar Current (ACC) is the strongest current system in the world oceans,
linking the Atlantic, Indian and Pacific basins.

Antarctic Circumpolar Current and Antarctic Convergence

The Antarctic Circumpolar Current moves perpetually eastward — chasing and joining itself, and
at 21,000 km (13,000 mi) in length — it comprises the world's longest ocean current, transporting 130 million cubic metres per second (4.6×10^9 cu ft/s) of water – 100 times the flow of all the
world's rivers.

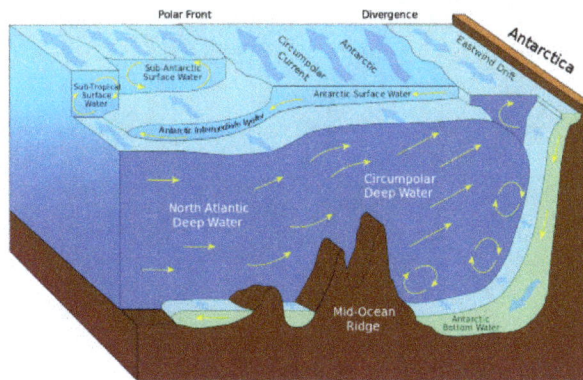

Upwelling in the Southern Ocean

Several processes operate along the coast of Antarctica to produce, in the Southern Ocean, types
of water masses not produced elsewhere in the oceans of the Southern Hemisphere. One of these
is the Antarctic Bottom Water, a very cold, highly saline, dense water that forms under sea ice.

Associated with the Circumpolar Current is the Antarctic Convergence encircling Antarctica, where
cold northward-flowing Antarctic waters meet the relatively warmer waters of the subantarctic,
Antarctic waters predominantly sink beneath subantarctic waters, while associated zones of mix-

ing and upwelling create a zone very high in nutrients. These nurture high levels of phytoplankton with associated copepods and Antarctic krill, and resultant foodchains supporting fish, whales, seals, penguins, albatrosses and a wealth of other species.

The Antarctic Convergence is considered to be the best natural definition of the northern extent of the Southern Ocean.

Upwelling

Large-scale upwelling is found in the Southern Ocean. Strong westerly (eastward) winds blow around Antarctica, driving a significant flow of water northwards. This is actually a type of coastal upwelling. Since there are no continents in a band of open latitudes between South America and the tip of the Antarctic Peninsula, some of this water is drawn up from great depths. In many numerical models and observational syntheses, the Southern Ocean upwelling represents the primary means by which deep dense water is brought to the surface. Shallower, wind-driven upwelling is also found off the west coasts of North and South America, northwest and southwest Africa, and southwest and southeast Australia, all associated with oceanic subtropical high pressure circulations.

Some models of the ocean circulation suggest that broad-scale upwelling occurs in the tropics, as pressure driven flows converge water toward the low latitudes where it is diffusively warmed from above. The required diffusion coefficients, however, appear to be larger than are observed in the real ocean. Nonetheless, some diffusive upwelling does probably occur.

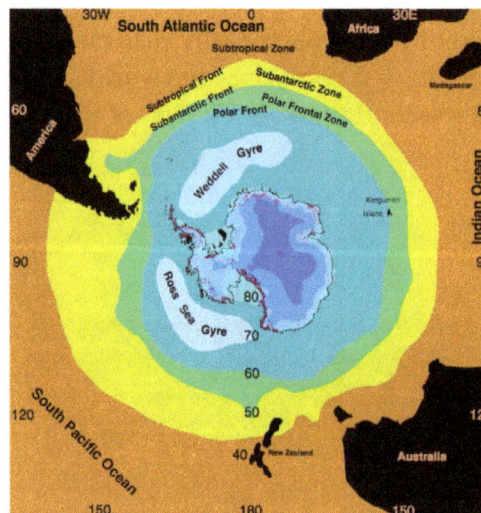

Location of the Southern Ocean gyres.

Ross and Weddell Gyres

The Ross Gyre and Weddell Gyre are two gyres that exist within the Southern Ocean. The gyres are located in the Ross Sea and Weddell Sea respectively, and both rotate clockwise. The gyres are formed by interactions between the Antarctic Circumpolar Current and the Antarctic Continental Shelf.

Sea ice has been noted to persist in the central area of the Ross Gyre. There is some evidence that global warming has resulted in some decrease of the salinity of the waters of the Ross Gyre since the 1950s.

Due to the Coriolis effect acting to the left in the Southern Hemisphere and the resulting Ekman transport away from the centres of the Weddell Gyre, these regions are very productive due to up-welling of cold, nutrient rich water.

Climate

Sea temperatures vary from about −2 to 10 °C (28 to 50 °F). Cyclonic storms travel eastward around the continent and frequently become intense because of the temperature contrast between ice and open ocean. The ocean-area from about latitude 40 south to the Antarctic Circle has the strongest average winds found anywhere on Earth. In winter the ocean freezes outward to 65 degrees south latitude in the Pacific sector and 55 degrees south latitude in the Atlantic sector, lowering surface temperatures well below 0 degrees Celsius. At some coastal points, however, persistent intense drainage winds from the interior keep the shoreline ice-free throughout the winter.

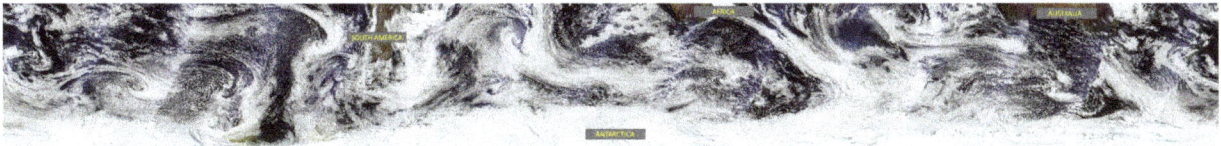

Clouds over Southern Ocean with Continent labels.

Biodiversity

Orca (*Orcinus orca*) hunting a Weddell seal in the Southern Ocean.

Animals

A variety of marine animals exist and rely, directly or indirectly, on the phytoplankton in the Southern Ocean. Antarctic sea life includes penguins, blue whales, orcas, colossal squids and fur seals. The emperor penguin is the only penguin that breeds during the winter in Antarctica, while the Adélie penguin breeds farther south than any other penguin. The rockhopper penguin has distinctive feathers around the eyes, giving the appearance of elaborate eyelashes. King penguins, chinstrap penguins, and gentoo penguins also breed in the Antarctic.

The Antarctic fur seal was very heavily hunted in the 18th and 19th centuries for its pelt by sealers from the United States and the United Kingdom. The Weddell seal, a "true seal", is named after Sir James Weddell, commander of British sealing expeditions in the Weddell Sea. Antarctic krill, which congregates in large schools, is the keystone species of the ecosystem of the Southern Ocean, and is an important food organism for whales, seals, leopard seals, fur seals, squid, icefish, penguins, albatrosses and many other birds.

The benthic communities of the seafloor are diverse and dense, with up to 155,000 animals found in 1 square metre (10.8 sq ft). As the seafloor environment is very similar all around the Antarctic, hundreds of species can be found all the way around the mainland, which is a uniquely wide distribution for such a large community. Deep-sea gigantism is common among these animals.

A census of sea life carried out during the International Polar Year and which involved some 500 researchers was released in 2010. The research is part of the global Census of Marine Life (CoML) and has disclosed some remarkable findings. More than 235 marine organisms live in both polar regions, having bridged the gap of 12,000 km (7,456 mi). Large animals such as some cetaceans and birds make the round trip annually. More surprising are small forms of life such as mudworms, sea cucumbers and free-swimming snails found in both polar oceans. Various factors may aid in their distribution – fairly uniform temperatures of the deep ocean at the poles and the equator which differ by no more than 5 °C, and the major current systems or marine conveyor belt which transport egg and larva stages.

A wandering albatross (*Diomedea exulans*) on South Georgia

Birds

The rocky shores of mainland Antarctica and its offshore islands provide nesting space for over 100 million birds every spring. These nesters include species of albatrosses, petrels, skuas, gulls and terns. The insectivorous South Georgia pipit is endemic to South Georgia and some smaller surrounding islands. Freshwater ducks inhabit South Georgia and the Kerguelen Islands.

The flightless penguins are all located in the Southern Hemisphere, with the greatest concentration located on and around Antarctica. Four of the 18 penguin species live and breed on the mainland and its close offshore islands. Another four species live on the subantarctic islands. Emperor penguins have four overlapping layers of feathers, keeping them warm. They are the only Antarctic animal to breed during the winter.

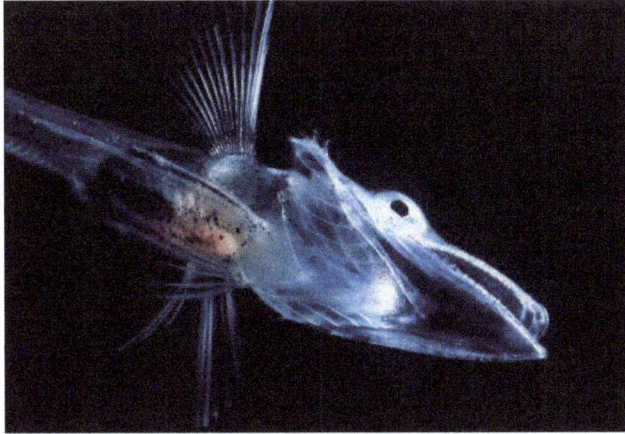

Fish of the Notothenioidei suborder, such as this young icefish, are mostly endemic to Antarctica.

Fish

There are very few species of fish in the Southern Ocean. The Channichthyidae family, also known as white-blooded fish, are only found in the Southern Ocean. They lack haemoglobin in their blood, resulting in their blood being colourless. One Channichthyidae species, the mackerel icefish (*Champsocephalus gunnari*), was once the most common fish in coastal waters less than 400 metres (1,312 ft) deep, but was overfished in the 1970s and 1980s. Schools of icefish spend the day at the seafloor and the night higher in the water column eating plankton and smaller fish.

There are two species from the *Dissostichus* genus, the Antarctic toothfish (*Dissostichus mawsoni*) and the Patagonian toothfish (*Dissostichus eleginoides*). These two species live on the seafloor 100–3,000 metres (328–9,843 ft) deep, and can grow to around 2 metres (7 ft) long weighing up to 100 kilograms (220 lb), living up to 45 years. The Antarctic toothfish lives close to the Antarctic mainland, whereas the Patagonian toothfish lives in the relatively warmer subantarctic waters. Due to the low water temperatures around the mainland, the Antarctic toothfish has antifreeze proteins in its blood and tissues. Toothfish are commercially fished, and illegal overfishing has reduced toothfish populations.

Another abundant fish group is the *Notothenia* genus, which like the Antarctic toothfish have antifreeze in their bodies.

Weddell seals (*Leptonychotes weddellii*) are the most southerly of Antarctic mammals.

Mammals

Seven pinniped species inhabit Antarctica. The largest, the elephant seal (*Mirounga leonina*), can reach up to 4,000 kilograms (8,818 lb), while females of the smallest, the Antarctic fur seal (*Arctocephalus gazella*), reach only 150 kilograms (331 lb). These two species live north of the sea ice, and breed in harems on beaches. The other four species can live on the sea ice. Crabeater seals (*Lobodon carcinophagus*) and Weddell seals (*Leptonychotes weddellii*) form breeding colonies, whereas leopard seals (*Hydrurga leptonyx*) and Ross seals (*Ommatophoca rossii*) live solitary lives. Although these species hunt underwater, they breed on land or ice and spend a great deal of time there, as they have no terrestrial predators.

The four species that inhabit sea ice are thought to make up 50% of the total biomass of the world's seals. Crabeater seals have a population of around 15 million, making them one of the most numerous large animals on the planet. The New Zealand sea lion (*Phocarctos hookeri*), one of the rarest and most localised pinnipeds, breeds almost exclusively on the subantarctic Auckland Islands, although historically it had a wider range. Out of all permanent mammalian residents, the Weddell seals live the furthest south.

There are 10 cetacean species found in the Southern Ocean; six baleen whales, and four toothed whales. The largest of these, the blue whale (*Balaenoptera musculus*), grows to 24 metres (79 ft) long weighing 84 tonnes. Many of these species are migratory, and travel to tropical waters during the Antarctic winter.

Antarctic krill (*Euphausia superba*) are a keystone species of the food web.

Arthropods

Five species of krill, small free-swimming crustaceans, are found in the Southern Ocean. The Antarctic krill (*Euphausia superba*) is one of the most abundant animal species on earth, with a biomass of around 500 million tonnes. Each individual is 6 centimetres (2.4 in) long and weighs over 1 gram (0.035 oz). The swarms that form can stretch for kilometres, with up to 30,000 individuals per 1 cubic metre (35 cu ft), turning the water red. Swarms usually remain in deep water during the day, ascending during the night to feed on plankton. Many larger animals depend on krill for their own survival. During the winter when food is scarce, adult Antarctic krill can revert to a smaller juvenile stage, using their own body as nutrition.

Many benthic crustaceans have a non-seasonal breeding cycle, and some raise their young in a brood pouch. *Glyptonotus antarcticus* is an unusually large benthic isopod, reaching 20 centimetres (8 in) in length weighing 70 grams (2.47 oz). Amphipods are abundant in soft sediments, eating a range of items, from algae to other animals.

Slow moving sea spiders are common, sometimes growing as large as a human hand. They feed on the corals, sponges, and bryozoans that litter the seabed.

A female warty squid (*Moroteuthis ingens*)

Invertebrates

Many aquatic molluscs are present in Antarctica. Bivalves such as *Adamussium colbecki* move around on the seafloor, while others such as *Laternula elliptica* live in burrows filtering the water above. There are around 70 cephalopod species in the Southern Ocean, the largest of which is the giant squid (*Architeuthis* sp.), which at 15 metres (49 ft) is the largest invertebrate in the world. Squid makes up the entire diet of some animals, such as grey-headed albatrosses and sperm whales, and the warty squid (*Moroteuthis ingens*) is one of the subantarctic's most preyed upon species by vertebrates.

The sea urchin genus *Abatus* burrow through the sediment eating the nutrients they find in it. Two species of salps are common in Antarctic waters, *Salpa thompsoni* and *Ihlea racovitzai*. *Salpa thompsoni* is found in ice-free areas, whereas *Ihlea racovitzai* is found in the high latitude areas near ice. Due to their low nutritional value, they are normally only eaten by fish, with larger animals such as birds and marine mammals only eating them when other food is scarce.

Antarctic sponges are long lived, and sensitive to environmental changes due to the specificity of the symbiotic microbial communities within them. As a result, they function as indicators of environmental health.

Environment

Current Issues

Increased solar ultraviolet radiation resulting from the Antarctic ozone hole has reduced marine primary productivity (phytoplankton) by as much as 15% and has started damaging the DNA of some fish. Illegal, unreported, and unregulated fishing, especially the landing of an estimated five to six times more Patagonian toothfish than the regulated fishery, likely affects the sustainability of the stock. Long-line fishing for toothfish causes a high incidence of seabird mortality.

International Agreements

All international agreements regarding the world's oceans apply to the Southern Ocean. In addition, it is subject to these agreements specific to the region:

- The *Southern Ocean Whale Sanctuary* of the International Whaling Commission (IWC) prohibits commercial whaling south of 40 degrees south (south of 60 degrees south between 50 degrees and 130 degrees west). Japan regularly does not recognize this provision, because the sanctuary violates IWC charter. Since the scope of the sanctuary is limited to commercial whaling, in regard to its whaling permit and whaling for scientific research, a Japanese fleet carried out an annual whale-hunt in the region. On 31 March 2014, the International Court of Justice ruled that Japan's whaling program, which Japan has long claimed is for scientific purposes, was a cloak for commercial whaling, and no further permits would be granted.

- *Convention for the Conservation of Antarctic Seals* is part of the *Antarctic Treaty System*. It was signed at the conclusion of a multilateral conference in London on 11 February 1972.

- *Convention for the Conservation of Antarctic Marine Living Resources* (CCAMLR) is part of the *Antarctic Treaty System*. The Convention was entered into force on 7 April 1982 and has its goal is to preserve marine life and environmental integrity in and near Antarctica. It was established in large part to concerns that an increase in krill catches in the Southern Ocean could have a serious impact on populations of other marine life which are dependent upon krill for food.

Many nations prohibit the exploration for and the exploitation of mineral resources south of the fluctuating Antarctic Convergence, which lies in the middle of the Antarctic Circumpolar Current and serves as the dividing line between the very cold polar surface waters to the south and the warmer waters to the north. The Antarctic Treaty covers the portion of the globe south of sixty degrees south, it prohibits new claims to Antarctica.

The *Convention for the Conservation of Antarctic Marine Living Resources* applies to the area south of 60° South latitude as well as the areas further north up to the limit of the Antarctic Convergence.

Economy

Between 1 July 1998 and 30 June 1999, fisheries landed 119,898 tonnes, of which 85% consisted of krill and 14% of Patagonian toothfish. International agreements came into force in late 1999 to reduce illegal, unreported, and unregulated fishing, which in the 1998–99 season landed five to six times more Patagonian toothfish than the regulated fishery.

Ports and Harbors

Major operational ports include: Rothera Station, Palmer Station, Villa Las Estrellas, Esperanza Base, Mawson Station, McMurdo Station, and offshore anchorages in Antarctica.

Few ports or harbors exist on the southern (Antarctic) coast of the Southern Ocean, since ice conditions limit use of most shores to short periods in midsummer; even then some require icebreaker escort for access. Most Antarctic ports are operated by government research stations and, except in an emergency, remain closed to commercial or private vessels; vessels in any port south of 60 degrees south are subject to inspection by Antarctic Treaty observers.

Severe cracks in an ice pier in use for four seasons at McMurdo Station slowed cargo operations in 1983 and proved a safety hazard.

The Southern Ocean's southernmost port operates at McMurdo Station at

77°50′S 166°40′E77.833°S 166.667°E. Winter Quarters Bay forms a small harbor, on the southern tip of Ross Island where a floating ice pier makes port operations possible in summer. Operation Deep Freeze personnel constructed the first ice pier at McMurdo in 1973.

Based on the original 1928 IHO delineation of the Southern Ocean (and the 1937 delineation if the Great Australian Bight is considered integral), Australian ports and harbors between Cape Leeuwin and Cape Otway on the Australian mainland and along the west coast of Tasmania would also be identified as ports and harbors existing in the Southern Ocean. These would include the larger ports and harbors of Albany, Thevenard, Port Lincoln, Whyalla, Port Augusta, Port Adelaide, Portland, Warrnambool, and Macquarie Harbour.

Yacht races have been held in the Southern Ocean, such as the Volvo Ocean Race, Velux 5 Oceans Race, Vendée Globe, Jules Verne Trophy and Global Challenge.

Pacific Ocean

The Pacific Ocean is the largest of the Earth's oceanic divisions. It extends from the Arctic Ocean in the north to the Southern Ocean (or, depending on definition, to Antarctica) in the south and is bounded by Asia and Australia in the west and the Americas in the east.

At 165.25 million square kilometers (63.8 million square miles) in area, this largest division of the World Ocean—and, in turn, the hydrosphere—covers about 46% of the Earth's water surface and about one-third of its total surface area, making it larger than all of the Earth's land area combined.

The equator subdivides it into the North Pacific Ocean and South Pacific Ocean, with two exceptions: the Galápagos and Gilbert Islands, while straddling the equator, are deemed wholly within the South Pacific. The Mariana Trench in the western North Pacific is the deepest point in the world, reaching a depth of 10,911 metres (35,797 ft).

Though the peoples of Asia and Oceania have travelled the Pacific Ocean since prehistoric times, the eastern Pacific was first sighted by Europeans in the early 16th century when Spanish explorer Vasco Núñez de Balboa crossed the Isthmus of Panama in 1513 and discovered the great "southern sea" which he named *Mar del Sur*. The ocean's current name was coined by Portuguese explorer Ferdinand Magellan during the Spanish circumnavigation of the world in 1521, as he encountered favourable winds on reaching the ocean. He called it *Mar Pacífico*, which in both Portuguese and Spanish means "peaceful sea".

History

Early Migrations

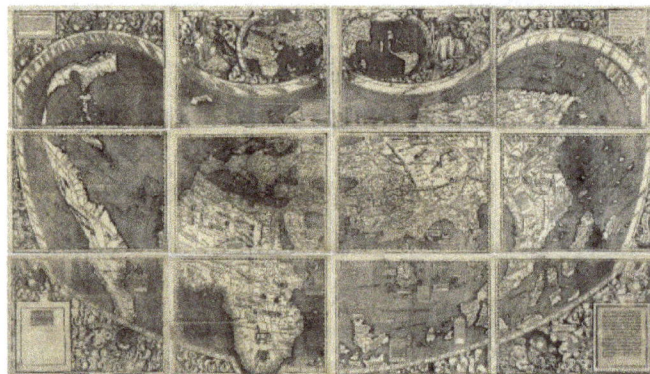

Universalis Cosmographia, the Waldseemüller map dated 1507, from a time when the nature of the Americas was ambiguous, particularly North America, as a possible part of Asia, was the first map to show the Americas separating two distinct oceans. South America was generally considered a "new world" and shows the name "America" for the first time, after Amerigo Vespucci

Made in 1529, the Diogo Ribeiro map was the first to show the Pacific at about its proper size

Important human migrations occurred in the Pacific in prehistoric times. About 3000 BC, the Austronesian peoples on the island of Taiwan mastered the art of long-distance canoe travel and spread themselves and their languages south to the Philippines, Indonesia, and maritime Southeast Asia; west towards Madagascar; southeast towards New Guinea and Melanesia (intermarrying with native Papuans); and east to the islands of Micronesia, Oceania and Polynesia.

Maris Pacifici by Ortelius (1589). One of the first printed maps to show the Pacific Ocean

USS *Lexington* under air attack on 8 May 1942 during the Battle of the Coral Sea

Long-distance trade developed all along the coast from Mozambique to Japan. Trade, and therefore knowledge, extended to the Indonesian islands but apparently not Australia. By at least 878 when there was a significant Islamic settlement in Canton much of this trade was controlled by Arabs or Muslims. In 219 BC Xu Fu sailed out into the Pacific searching for the elixir of immortality. From 1404 to 1433 Zheng He led expeditions into the Indian Ocean.

European Exploration

Map of the Pacific Ocean during European Exploration, circa 1702–1707.

Map of the Pacific Ocean during European Exploration, circa 1754.

The first contact of European navigators with the western edge of the Pacific Ocean was made by the Portuguese expeditions of António de Abreu and Francisco Serrão to the Maluku Islands in 1512, and with Jorge Álvares's expedition to southern China in 1513, both ordered by Afonso de Albuquerque.

The east side of the ocean was discovered by Spanish explorer Vasco Núñez de Balboa in 1513 after his expedition crossed the Isthmus of Panama and reached a new ocean. He named it *Mar del Sur* (literally, "Sea of the South" or "South Sea") because the ocean was to the south of the coast of the isthmus where he first observed the Pacific.

Later, Portuguese explorer Ferdinand Magellan sailed the Pacific on a Castilian (*Spanish*) expedition of world circumnavigation starting in 1519. Magellan called the ocean *Pacífico* (or "Pacific" meaning, "peaceful") because, after sailing through the stormy seas off Cape Horn, the expedition found calm waters. The ocean was often called the *Sea of Magellan* in his honor until the eighteenth century. Although Magellan himself died in the Philippines in 1521, Spanish Basque navigator Juan Sebastián Elcano led the expedition back to Spain across the Indian Ocean and round the Cape of Good Hope, completing the first world circumnavigation in 1522. Sailing around and east of the Moluccas, between 1525 and 1527, Portuguese expeditions discovered the Caroline Islands and Papua New Guinea. In 1542–43 the Portuguese also reached Japan.

In 1564, five Spanish ships consisting of 379 explorers crossed the ocean from Mexico led by Miguel López de Legazpi and sailed to the Philippines and Mariana Islands. For the remainder of the 16th century, Spanish influence was paramount, with ships sailing from Mexico and Peru across the Pacific Ocean to the Philippines, via Guam, and establishing the Spanish East Indies. The Manila galleons operated for two and a half centuries linking Manila and Acapulco, in one of the longest trade routes in history. Spanish expeditions also discovered Tuvalu, the Marquesas, the Cook Islands, the Solomon Islands, and the Admiralty Islands in the South Pacific.

Later, in the quest for Terra Australis (i.e., "the [great] Southern Land"), Spanish explorers in the 17th century discovered the Pitcairn and Vanuatu archipelagos, and sailed the Torres Strait between Australia and New Guinea, named after navigator Luís Vaz de Torres. Dutch explorers, sailing around southern Africa, also engaged in discovery and trade; Abel Janszoon Tasman discovered Tasmania and New Zealand in 1642.

In the 16th and 17th century Spain considered the Pacific Ocean a *Mare clausum*—a sea closed to other naval powers. As the only known entrance from the Atlantic the Strait of Magellan was at times patrolled by fleets sent to prevent entrance of non-Spanish ships. On the western end of the Pacific Ocean the Dutch threatened the Spanish Philippines.

The 18th century marked the beginning of major exploration by the Russians in Alaska and the Aleutian Islands. Spain also sent expeditions to the Pacific Northwest reaching Vancouver Island in southern Canada, and Alaska. The French explored and settled Polynesia, and the British made three voyages with James Cook to the South Pacific and Australia, Hawaii, and the North American Pacific Northwest. In 1768, Pierre-Antoine Véron, a young astronomer accompanying Louis Antoine de Bougainville on his voyage of exploration, established the width of the Pacific with precision for the first time in history. One of the earliest voyages of scientific exploration was organized by Spain in the Malaspina Expedition of 1789–1794. It sailed vast areas of the Pacific, from Cape Horn to Alaska, Guam and the Philippines, New Zealand, Australia, and the South Pacific.

New Imperialism

The Bathyscaphe Trieste, before her record dive to the bottom of the Mariana Trench, 23 January 1960

Growing imperialism during the 19th century resulted in the occupation of much of Oceania by other European powers, and later, Japan and the United States. Significant contributions to oceanographic knowledge were made by the voyages of HMS *Beagle* in the 1830s, with Charles Darwin aboard; HMS *Challenger* during the 1870s; the USS *Tuscarora* (1873–76); and the German *Gazelle* (1874–76).

Dupetit Thouars taking over Tahiti on 9 September 1842.

In Oceania, France got a leading position as imperial power after making Tahiti and New Caledonia protectorates in 1842 and 1853 respectively. After navy visits to Easter Island in 1875 and 1887, Chilean navy officer Policarpo Toro managed to negotiate an incorporation of the island into Chile with native Rapanui in 1888. By occupying Easter Island, Chile joined the imperial nations. By 1900 nearly all Pacific islands were in control of Britain, France, United States, Germany, Japan, and Chile.

Although the United States gained control of Guam and the Philippines from Spain in 1898, Japan controlled most of the western Pacific by 1914 and occupied many other islands during World War II. However, by the end of that war, Japan was defeated and the U.S. Pacific Fleet was the virtual master of the ocean. Since the end of World War II, many former colonies in the Pacific have become independent states.

Geography

The Pacific separates Asia and Australia from the Americas. It may be further subdivided by the equator into northern (North Pacific) and southern (South Pacific) portions. It extends from the Antarctic region in the South to the Arctic in the north. The Pacific Ocean encompasses approximately one-third of the Earth's surface, having an area of 165.2 million square kilometers (63.8 million square miles)—significantly larger than Earth's entire landmass of some 150 million square kilometers (58 million square miles).

Sunset over the Pacific Ocean as seen from the International Space Station. Anvil tops of thunderclouds are also visible.

South Pacific Ocean from Newcastle NSW Shore

Extending approximately 15,500 km (9,600 mi) from the Bering Sea in the Arctic to the northern extent of the circumpolar Southern Ocean at 60°S (older definitions extend it to Antarctica's Ross Sea), the Pacific reaches its greatest east-west width at about 5°N latitude, where it stretches approximately 19,800 km (12,300 mi) from Indonesia to the coast of Colombia—halfway around the world, and more than five times the diameter of the Moon. The lowest known point on Earth—the Mariana Trench—lies 10,911 m (35,797 ft; 5,966 fathoms) below sea level. Its average depth is 4,280 m (14,040 ft; 2,340 fathoms), putting the total water volume at 710,000,000 cubic kilometers.

Due to the effects of plate tectonics, the Pacific Ocean is currently shrinking by roughly 2.5 centimetres (0.98 in) per year on three sides, roughly averaging 0.52 square kilometres (0.20 sq mi) a year. By contrast, the Atlantic Ocean is increasing in size.

Along the Pacific Ocean's irregular western margins lie many seas, the largest of which are the Celebes Sea, Coral Sea, East China Sea, Philippine Sea, Sea of Japan, South China Sea, Sulu Sea, Tasman Sea, and Yellow Sea. The Indonesian Seaway (including the Strait of Malacca and Torres Strait) joins the Pacific and the Indian Ocean to the west, and Drake Passage and the Strait of Magellan link the Pacific with the Atlantic Ocean on the east. To the north, the Bering Strait connects the Pacific with the Arctic Ocean.

Storm in Pacifica, California

As the Pacific straddles the 180th meridian, the *West Pacific* (or *western Pacific*, near Asia) is in the Eastern Hemisphere, while the *East Pacific* (or *eastern Pacific*, near the Americas) is in the Western Hemisphere.

The Southern Pacific Ocean harbors the Southeast Indian Ridge crossing from south of Australia turning into the Pacific-Antarctic Ridge (north of the South Pole) and merges with another ridge (south of South American) to form the East Pacific Rise which also connects with another ridge (south of North America) which overlooks the Juan de Fuca Ridge.

For most of Magellan's voyage from the Strait of Magellan to the Philippines, the explorer indeed found the ocean peaceful. However, the Pacific is not always peaceful. Many tropical storms batter the islands of the Pacific. The lands around the Pacific Rim are full of volcanoes and often affected by earthquakes. Tsunamis, caused by underwater earthquakes, have devastated many islands and in some cases destroyed entire towns.

The Martin Waldseemüller map of 1507 was the first to show the Americas separating two distinct oceans. Later, the Diogo Ribeiro map of 1529 was the first to show the Pacific at about its proper size.

Bordering Countries and Territories

Sovereign Nations

- 🇦🇺 Australia

- Brunei
- Cambodia
- Canada
- Chile
- China[1]
- Colombia
- Costa Rica
- Ecuador
- El Salvador
- Federated States of Micronesia
- Fiji
- Guatemala
- Honduras
- Indonesia
- Japan
- Kiribati
- North Korea
- South Korea
- Malaysia
- Marshall Islands
- Mexico
- Nauru
- New Zealand
- Nicaragua
- Palau
- Panama
- Papua New Guinea
- Peru

- Philippines
- Russia
- Samoa
- Singapore
- Solomon Islands
- Taiwan[1]
- Thailand
- Timor-Leste
- Tonga
- Tuvalu
- United States
- Vanuatu
- Vietnam

[1] *The status of Taiwan and China is disputed.*

Territories

- American Samoa (US)
- Baker Island (US)
- Cook Islands (New Zealand)
- Coral Sea Islands (Australia)
- Easter Island (Chile)
- French Polynesia (France)
- Guam (US)
- Hong Kong (China)
- Howland Island (US)
- Jarvis Island (US)
- Johnston Island (US)
- Kingman Reef (US)

- Macau (China)
- Midway Atoll (US)
- New Caledonia (France)
- Niue (New Zealand)
- Norfolk Island (Australia)
- Northern Mariana Islands (US)
- Palmyra Atoll (US)
- Pitcairn Islands (UK)
- Tokelau (New Zealand)
- Wallis and Futuna (France)
- Wake Island (US)

Landmasses and Islands

The islands entirely within the Pacific Ocean can be divided into three main groups known as Micronesia, Melanesia and Polynesia. Micronesia, which lies north of the equator and west of the International Date Line, includes the Mariana Islands in the northwest, the Caroline Islands in the center, the Marshall Islands to the west and the islands of Kiribati in the southwest.

Melanesia, to the southwest, includes New Guinea, the world's second largest island after Greenland and by far the largest of the Pacific islands. The other main Melanesian groups from north to south are the Bismarck Archipelago, the Solomon Islands, Santa Cruz, Vanuatu, Fiji and New Caledonia.

The largest area, Polynesia, stretching from Hawaii in the north to New Zealand in the south, also encompasses Tuvalu, Tokelau, Samoa, Tonga and the Kermadec Islands to the west, the Cook Islands, Society Islands and Austral Islands in the center, and the Marquesas Islands, Tuamotu, Mangareva Islands and Easter Island to the east.

Islands in the Pacific Ocean are of four basic types: continental islands, high islands, coral reefs and uplifted coral platforms. Continental islands lie outside the andesite line and include New Guinea, the islands of New Zealand, and the Philippines. Some of these islands are structurally associated with nearby continents. High islands are of volcanic origin, and many contain active volcanoes. Among these are Bougainville, Hawaii, and the Solomon Islands.

The coral reefs of the South Pacific are low-lying structures that have built up on basaltic lava flows under the ocean's surface. One of the most dramatic is the Great Barrier Reef off northeastern Australia with chains of reef patches. A second island type formed of coral is the uplifted coral platform, which is usually slightly larger than the low coral islands. Examples include Banaba (formerly Ocean Island) and Makatea in the Tuamotu group of French Polynesia.

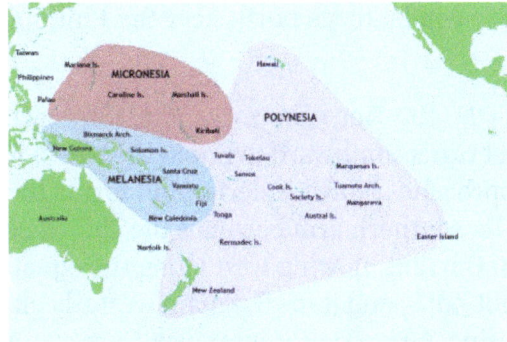

Micronesia, Melanesia, and Polynesia

Tahuna maru islet, French Polynesia

Water Characteristics

Sunset in Monterey County, California, U.S.

The volume of the Pacific Ocean, representing about 50.1 percent of the world's oceanic water, has been estimated at some 714 million cubic kilometers. Surface water temperatures in the Pacific can vary from −1.4 °C (29.5 °F), the freezing point of sea water, in the poleward areas to about 30 °C (86 °F) near the equator. Salinity also varies latitudinally, reaching a maximum of 37 parts per thousand in the southeastern area. The water near the equator, which can have a salinity as low as 34 parts per thousand, is less salty than that found in the mid-latitudes because of abundant equatorial precipitation throughout the year. The lowest counts of less than 32 parts per thousand are found in the far north as less evaporation of seawater takes place in these frigid areas. The motion of Pacific waters is generally clockwise in the Northern Hemisphere (the North Pacific gyre) and counter-clockwise in the Southern Hemisphere. The North Equatorial Current, driven westward along latitude 15°N by the trade winds, turns north near the Philippines to become the warm Japan or Kuroshio Current.

Turning eastward at about 45°N, the Kuroshio forks and some water moves northward as the Aleutian Current, while the rest turns southward to rejoin the North Equatorial Current. The Aleutian Current branches as it approaches North America and forms the base of a counter-clockwise circulation in the Bering Sea. Its southern arm becomes the chilled slow, south-flowing California Current. The South Equatorial Current, flowing west along the equator, swings southward east of New Guinea, turns east at about 50°S, and joins the main westerly circulation of the South Pacific, which includes the Earth-circling Antarctic Circumpolar Current. As it approaches the Chilean coast, the South Equatorial Current divides; one branch flows around Cape Horn and the other turns north to form the Peru or Humboldt Current.

Climate

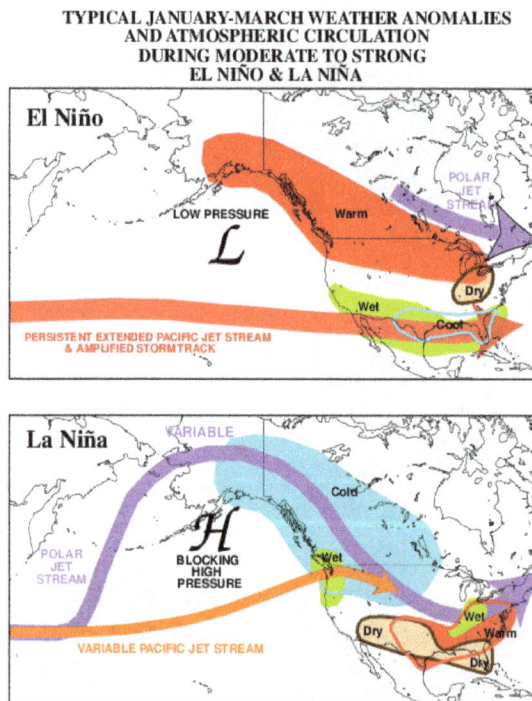

Impact of El Niño and La Niña on North America

The climate patterns of the Northern and Southern Hemispheres generally mirror each other. The trade winds in the southern and eastern Pacific are remarkably steady while conditions in the North Pacific are far more varied with, for example, cold winter temperatures on the east coast of Russia contrasting with the milder weather off British Columbia during the winter months due to the preferred flow of ocean currents.

In the tropical and subtropical Pacific, the El Niño Southern Oscillation (ENSO) affects weather conditions. To determine the phase of ENSO, the most recent three-month sea surface temperature average for the area approximately 3,000 kilometres (1,900 mi) to the southeast of Hawaii is computed, and if the region is more than 0.5 °C (0.9 °F) above or below normal for that period, then an El Niño or La Niña is considered in progress.

Typhoon Tip at global peak intensity on 12 October 1979

In the tropical western Pacific, the monsoon and the related wet season during the summer months contrast with dry winds in the winter which blow over the ocean from the Asian landmass. World-wide, tropical cyclone activity peaks in late summer, when the difference between temperatures aloft and sea surface temperatures is the greatest. However, each particular basin has its own seasonal patterns. On a worldwide scale, May is the least active month, while September is the most active month. November is the only month in which all the tropical cyclone basins are active. Cyclones are liable to form south of Mexico, striking the western Mexican coast and occasionally the southwestern United States between June and October, with those forming in the western Pacific moving into southeast and east Asia from May to December.

In the arctic, icing from October to May can present a hazard for shipping while persistent fog occurs from June to December. A climatological low in the Gulf of Alaska keeps the southern coast wet and mild during the winter months. The Westerlies and associated jet stream within the Mid-Latitudes can be particularly strong, especially in the Southern Hemisphere, due to the temperature difference between the tropics and Antarctica, which records the coldest temperature readings on the planet. In the Southern hemisphere, because of the stormy and cloudy conditions associated with extratropical cyclones riding the jet stream, it is usual to refer to the Westerlies as the Roaring Forties, Furious Fifties and Shrieking Sixties according to the varying degrees of latitude.

Geology

The ocean was first mapped by Abraham Ortelius; he called it Maris Pacifici following Ferdinand Magellan's description of it as "a pacific sea" during his circumnavigation from 1519 to 1522. To Magellan, it seemed much more calm (pacific) than the Atlantic.

The andesite line is the most significant regional distinction in the Pacific. A petrologic boundary, it separates the deeper, mafic igneous rock of the Central Pacific Basin from the partially submerged continental areas of felsic igneous rock on its margins. The andesite line follows the western edge of the islands off California and passes south of the Aleutian arc, along the eastern edge of the Kamchatka Peninsula, the Kuril Islands, Japan, the Mariana Islands, the Solomon Islands, and New Zealand's North Island.

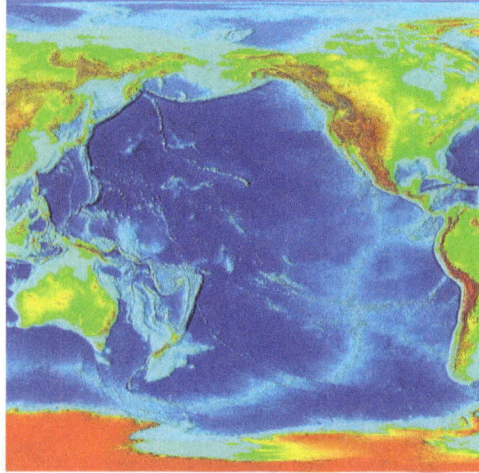

The Pacific is ringed by many volcanoes and oceanic trenches.

Ulawun stratovolcano situated on the island of New Britain, Papua New Guinea

The dissimilarity continues northeastward along the western edge of the Andes Cordillera along South America to Mexico, returning then to the islands off California. Indonesia, the Philippines, Japan, New Guinea, and New Zealand lie outside the andesite line.

Within the closed loop of the andesite line are most of the deep troughs, submerged volcanic mountains, and oceanic volcanic islands that characterize the Pacific basin. Here basaltic lavas gently flow out of rifts to build huge dome-shaped volcanic mountains whose eroded summits form island arcs, chains, and clusters. Outside the andesite line, volcanism is of the explosive type, and the Pacific Ring of Fire is the world's foremost belt of explosive volcanism. The Ring of Fire is named after the several hundred active volcanoes that sit above the various subduction zones.

The Pacific Ocean is the only ocean which is almost totally bounded by subduction zones. Only the Antarctic and Australian coasts have no nearby subduction zones.

Geological History

The Pacific Ocean was born 750 million years ago at the breakup of Rodinia, although it is generally called the Panthalassic Ocean until the breakup of Pangea, about 200 million years ago. The oldest Pacific Ocean floor is only around 180 Ma old, with older crust subducted by now.

Seamount Chains

The Pacific Ocean contains several long seamount chains, formed by hotspot volcanism. These include the Hawaiian–Emperor seamount chain and the Louisville seamount chain.

Economy

The exploitation of the Pacific's mineral wealth is hampered by the ocean's great depths. In shallow waters of the continental shelves off the coasts of Australia and New Zealand, petroleum and natural gas are extracted, and pearls are harvested along the coasts of Australia, Japan, Papua New Guinea, Nicaragua, Panama, and the Philippines, although in sharply declining volume in some cases.

Fishing

Fish are an important economic asset in the Pacific. The shallower shoreline waters of the continents and the more temperate islands yield herring, salmon, sardines, snapper, swordfish, and tuna, as well as shellfish. Overfishing has become a serious problem in some areas. For example, catches in the rich fishing grounds of the Okhotsk Sea off the Russian coast have been reduced by at least half since the 1990s as a result of overfishing.

Environmental Issues

Marine debris on a Hawaiian coast

The quantity of small plastic fragments floating in the north-east Pacific Ocean increased a hundredfold between 1972 and 2012.

Marine pollution is a generic term for the harmful entry into the ocean of chemicals or particles. The main culprits are those using the rivers for disposing of their waste. The rivers then empty into the ocean, often also bringing chemicals used as fertilizers in agriculture. The excess of oxygen-depleting chemicals in the water leads to hypoxia and the creation of a dead zone.

Marine debris, also known as marine litter, is human-created waste that has ended up floating in a lake, sea, ocean, or waterway. Oceanic debris tends to accumulate at the center of gyres and coastlines, frequently washing aground where it is known as beach litter.

In addition, the Pacific Ocean has served as the crash site of satellites, including Mars 96, Fobos-Grunt, and Upper Atmosphere Research Satellite.

Indian Ocean

The **Indian Ocean** is the third largest of the world's oceanic divisions, covering 70,560,000 km² (27,240,000 sq mi) (approximately 20% of the water on the Earth's surface). It is bounded by Asia on the north, on the west by Africa, on the east by Australia, and on the south by the Southern Ocean or, depending on definition, by Antarctica. It is named after the country of India. The Indian Ocean is known as *Ratnākara* (Sanskrit: रत्नाकर), "*the mine of gems*" in ancient Sanskrit literature, and as *Hind Mahāsāgar* (Hindi: हिन्द महासागर), "*the great Indian sea*", in Hindi.

Geography

A -17th century- 1658 Naval Map by Janssonius depicting the Indian Ocean, India and Arabia.

The borders of the Indian Ocean, as delineated by the International Hydrographic Organization in 1953 included the Southern Ocean but not the marginal seas along the northern rim, but in 2000 the IHO delimited the Southern Ocean separately, which removed waters south of 60°S from the Indian Ocean, but included the northern marginal seas. Meridionally, the Indian Ocean is delimited from the Atlantic Ocean by the 20° east meridian, running south from Cape Agulhas, and from the Pacific Ocean by the meridian of 146°55'E, running south from the southernmost point of Tasmania. The northernmost extent of the Indian Ocean is approximately 30° north in the Persian Gulf.

The ocean covers 70,560,000 km² (27,240,000 sq mi), including the Red Sea and the Persian Gulf but excluding the Southern Ocean, or 19.5% world's oceans; its volume is 264,000,000 km³ (63,000,000 cu mi) or 19.8% of oceans volume; it has an average depth of 3,741 m (12,274 ft) and a maximum depth of 7,906 m (25,938 ft).

The ocean's continental shelves are narrow, averaging 200 kilometres (120 mi) in width. An exception is found off Australia's western coast, where the shelf width exceeds 1,000 kilometres (620 mi). The average depth of the ocean is 3,890 m (12,762 ft). Its deepest point is Diamantina Deep in Diamantina Trench, at 8,047 m (26,401 ft) deep; also sometimes considered is Sunda Trench, at a depth of 7,258–7,725 m (23,812–25,344 ft). North of 50° south latitude, 86% of the main basin is covered by pelagic sediments, of which more than half is globigerina ooze. The remaining 14% is layered with terrigenous sediments. Glacial outwash dominates the extreme southern latitudes.

The major choke points include Bab el Mandeb, Strait of Hormuz, the Lombok Strait, the Strait of Malacca and the Palk Strait. Seas include the Gulf of Aden, Andaman Sea, Arabian Sea, Bay of Bengal, Great Australian Bight, Laccadive Sea, Gulf of Mannar, Mozambique Channel, Gulf of Oman, Persian Gulf, Red Sea and other tributary water bodies. The Indian Ocean is artificially connected to the Mediterranean Sea through the Suez Canal, which is accessible via the Red Sea.

Marginal Seas

Marginal seas, gulfs, bays and straits of the Indian Ocean include:

· Andaman Sea	· Gulf of Kutch	· Palk Strait connecting Arabian Sea and Bay of Bengal
· Arabian Sea	· Gulf of Khambat	
· Bay of Bengal	· Gulf of Oman	· Persian Gulf
· Great Australian Bight	· Indonesian Seaway (including the Malacca, Sunda and Torres Straits)	· Red Sea
· Gulf of Mannar		· Strait of Bab-el-Mandeb connecting Arabian Sea
· Gulf of Aden	· Laccadive Sea	
· Gulf of Carpentaria	· Mozambique Channel	

Climate

The climate north of the equator is affected by a monsoon climate. Strong north-east winds blow from October until April; from May until October south and west winds prevail. In the Arabian Sea the violent Monsoon brings rain to the Indian subcontinent. In the southern hemisphere, the winds are generally milder, but summer storms near Mauritius can be severe. When the monsoon winds change, cyclones sometimes strike the shores of the Arabian Sea and the Bay of Bengal. The Indian Ocean is the warmest ocean in the world.

Oceanography

Among the few large rivers flowing into the Indian Ocean are the Zambezi, Shatt al-Arab, Indus, Godavari, Krishna, Narmada, Ganges, Brahmaputra, Jubba and Irrawaddy River. The ocean's currents are mainly controlled by the monsoon. Two large gyres, one in the northern hemisphere flowing clockwise and one south of the equator moving anticlockwise (including the Agulhas Current and Agulhas Return Current), constitute the dominant flow pattern. During the winter monsoon, however, currents in the north are reversed.

Deep water circulation is controlled primarily by inflows from the Atlantic Ocean, the Red Sea, and Antarctic currents. North of 20° south latitude the minimum surface temperature is 22 °C (72 °F), exceeding 28 °C (82 °F) to the east. Southward of 40° south latitude, temperatures drop quickly.

Precipitation and evaporation leads to salinity variation in all oceans, and in the Indian Ocean salinity variations are driven by: (1) river inflow mainly from the Bay of Bengal, (2) fresher water from the Indonesian Throughflow; and (3) saltier water from the Red Sea and Persian Gulf. Surface water salinity ranges from 32 to 37 parts per 1000, the highest occurring in the Arabian Sea and in a belt between southern Africa and south-western Australia. Pack ice and icebergs are found throughout the year south of about 65° south latitude. The average northern limit of icebergs is 45° south latitude.

Geology

As the youngest of the major oceans, the Indian Ocean has active spreading ridges that are part of the worldwide system of mid-ocean ridges. In the Indian Ocean these spreading ridges meet at the Rodrigues Triple Point with the Central Indian Ridge, including the Carlsberg Ridge, separating the African Plate from the Indian Plate; the Southwest Indian Ridge separating the African Plate form the Antarctic Plate; and the Southeast Indian Ridge separating the Australian Plate from the Antarctic Plate.The Central Ridge runs north on the in-between across of the Arabian Peninsula and Africa into the Mediterranean Sea.

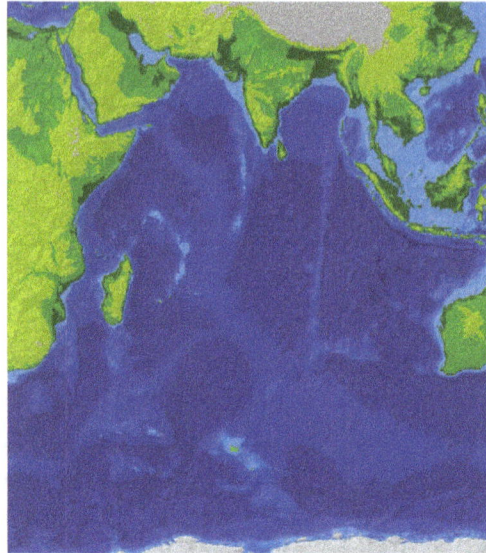

Bathymetric map of the Indian Ocean

A series of ridges and seamount chains produced by hotspots pass over the Indian Ocean. The Réunion hotspot (active 70-40 Ma) connects Réunion and the Mascarene Plateau to the Chagos-Laccadive Ridge and the Deccan Traps in north-western India; the Kerguelen hotspot (100-35 Ma) connects the Kerguelen Islands and Kerguelen Plateau to the Ninety East Ridge and the Rajmahal Traps in north-eastern India; the Marion hotspot (100-70 Ma) possibly connects Prince Edward Islands to the Eighty Five East Ridge. It should be noted that these hotspot tracks have been broken by the still active spreading ridges mentioned above.

Marine Life

Among the tropical oceans, the western Indian Ocean hosts one of the largest concentration of phytoplankton blooms in summer, due to the strong monsoon winds. The monsoonal wind forcing leads to a strong coastal and open ocean upwelling, which introduces nutrients into the upper zones where sufficient light is available for photosynthesis and phytoplankton production. These phytoplankton blooms support the marine ecosystem, as the base of the marine food web, and eventually the larger fish species. The Indian Ocean accounts for the second largest share of the most economically valuable tuna catch. Its fish are of great and growing importance to the bordering countries for domestic consumption and export. Fishing fleets from Russia, Japan, South Korea, and Taiwan also exploit the Indian Ocean, mainly for shrimp and tuna.

Research indicates that increasing ocean temperatures are taking a toll on the marine ecosystem. A study on the phytoplankton changes in the Indian Ocean indicates a decline of up to 20% in the marine phytoplankton in the Indian Ocean, during the past six decades. The tuna catch rates have also declined abruptly during the past half century, mostly due to increased industrial fisheries, with the ocean warming adding further stress to the fish species.

Endangered marine species include the dugong, seals, turtles, and whales.

An Indian Ocean garbage patch was discovered in 2010 covering at least 5 million square kilometres (1.9×10^6 sq mi). Riding the southern Indian Ocean Gyre, this vortex of plastic garbage constantly circulates the ocean from Australia to Africa, down the Mozambique Channel, and back to Australia in a period of six years, except for debris that get indefinitely stuck in the centre of the gyre.

History

The economically important Silk Road (red) and spice trade routes (blue) were blocked by the Ottoman Empire in ca. 1453 with the fall of the Byzantine Empire. This spurred exploration, and a new sea route around Africa was found, triggering the Age of Discovery.

First Settlements

The history of the Indian Ocean is marked by maritime trade; cultural and commercial exchange probably date back at least seven thousand years. During this period, independent, short-distance oversea communications along its littoral margins have evolved into an all-embracing network. The début of this network was not the achievement of a centralised or advanced civilisation but of local and regional exchange in the Persian Gulf, the Red Sea, and Arabian Sea. Sherds of Ubaid (2500-500 BCE) pottery have been found in the western Gulf at Dilmun, present-day Bahrain; traces of exchange between this trading centre and Mesopotamia. Sumerian traded grain, pottery, and bitumen (used for reed boats) for copper, stone, timber, tin, dates, onions, and pearls. Coast-bound vessels transported goods between the Harappa civilisation (2600–1900 BCE) in India (modern-day Pakistan and Gujarat in India) and the Persian Gulf and Egypt.

Periplus of the Erythraean Sea, an Alexandrian guide to the world beyond the Red Sea — including Africa and India — from the first century CE, not only gives insights into trade in the region but also shows that Roman and Greek sailors had already gained knowledge about the monsoon winds. The contemporaneous settlement of Madagascar by Indonesian sailors shows that the littoral margins of the Indian Ocean were being both well-populated and regularly traversed at least by this time. Albeit the monsoon must have been common knowledge in the Indian Ocean for centuries.

The world's earliest civilizations in Mesopotamia (beginning with Sumer), ancient Egypt, and the

Indian subcontinent (beginning with the Indus Valley civilization), which began along the valleys of the Tigris-Euphrates, Nile and Indus rivers respectively, all developed around the Indian Ocean. Civilizations soon arose in Persia (beginning with Elam) and later in Southeast Asia (beginning with Funan).

During Egypt's first dynasty (c. 3000 BC), sailors were sent out onto its waters, journeying to Punt, thought to be part of present-day Somalia. Returning ships brought gold and myrrh. The earliest known maritime trade between Mesopotamia and the Indus Valley (c. 2500 BC) was conducted along the Indian Ocean. Phoenicians of the late 3rd millennium BC may have entered the area, but no settlements resulted.

The Indian Ocean's relatively calmer waters opened the areas bordering it to trade earlier than the Atlantic or Pacific oceans. The powerful monsoons also meant ships could easily sail west early in the season, then wait a few months and return eastwards. This allowed ancient Indonesian peoples to cross the Indian Ocean to settle in Madagascar around 2000 BP.

Era of Discovery

In the 2nd or 1st century BC, Eudoxus of Cyzicus was the first Greek to cross the Indian Ocean. The probably fictitious sailor Hippalus is said to have discovered the direct route from Arabia to India around this time. During the 1st and 2nd centuries AD intensive trade relations developed between Roman Egypt and the Tamil kingdoms of the Cheras, Cholas and Pandyas in Southern India. Like the Indonesian peoples above, the western sailors used the monsoon to cross the ocean. The unknown author of the *Periplus of the Erythraean Sea* describes this route, as well as the commodities that were traded along various commercial ports on the coasts of the Horn of Africa and India circa 1 AD. Among these trading settlements were Mosylon and Opone on the Red Sea littoral.

British heavy cruisers *Dorsetshire* and *Cornwall* under Japanese air attack and heavily damaged on 5 April 1942

Unlike the Pacific Ocean where the civilization of the Polynesians reached most of the far flung islands and atolls and populated them, almost all the islands, archipelagos and atolls of the Indian Ocean were uninhabited until colonial times. Although there were numerous ancient civilizations in the coastal states of Asia and parts of Africa, the Maldives were the only island group in the Central Indian Ocean region where an ancient civilization flourished. Maldivian ships used the Indian Monsoon Current to travel to the nearby coasts.

From 1405 to 1433, Admiral Zheng He led large fleets of the Ming Dynasty on several treasure voyages through the Indian Ocean, ultimately reaching the coastal countries of East Africa.

In 1497, Portuguese navigator Vasco da Gama rounded the Cape of Good Hope and became the first European to sail to India and later the Far East. The European ships, armed with heavy cannon, quickly dominated trade. Portugal achieved pre-eminence by setting up forts at the important straits and ports. Their hegemony along the coasts of Africa and Asia lasted until the mid 17th century. Later, the Portuguese were challenged by other European powers. The Dutch East India Company (1602–1798) sought control of trade with the East across the Indian Ocean. France and Britain established trade companies for the area. From 1565, Spain established a major trading operation with the Manila Galleons in the Philippines and the Pacific. Spanish trading ships purposely avoided the Indian Ocean, following the Treaty of Tordesillas with Portugal. By 1815, Britain became the principal power in the Indian Ocean.

Industrial Era

The opening of the Suez Canal in 1869 revived European interest in the East, but no nation was successful in establishing trade dominance. Since World War II the United Kingdom was forced to withdraw from the area, to be replaced by India, the USSR, and the United States. The last two tried to establish hegemony by negotiating for naval base sites. Developing countries bordering the ocean, however, seek to have it made a "zone of peace" so that they may use its shipping lanes freely. The United Kingdom and United States maintain a military base on Diego Garcia atoll in the middle of the Indian Ocean.

Contemporary Era

On 26 December 2004, the countries surrounding the Indian Ocean were hit by a tsunami caused by the 2004 Indian Ocean earthquake. The waves resulted in more than 226,000 deaths and over 1 million people were left homeless.

In the late 2000s, the ocean evolved into a hub of pirate activity. By 2013, attacks off the Horn region's coast had steadily declined due to active private security and international navy patrols, especially by the Indian Navy.

Trade

The Indian Ocean provides major sea routes connecting the Middle East, Africa, and East Asia with Europe and the Americas. It carries a particularly heavy traffic of petroleum and petroleum products from the oil fields of the Persian Gulf and Indonesia. Large reserves of hydrocarbons are being tapped in the offshore areas of Saudi Arabia, Iran, India, and Western Australia. An estimated 40% of the world's offshore oil production comes from the Indian Ocean. Beach sands rich in heavy minerals, and offshore placer deposits are actively exploited by bordering countries, particularly India, Pakistan, South Africa, Indonesia, Sri Lanka, and Thailand.

A dhow off the coast of Kenya

Major Ports and Harbours

The Port of Singapore is the busiest port in the Indian Ocean, located in the Strait of Malacca where it meets the Pacific. Mumbai, Chennai, Kolkata, Kochi, Mormugao Port, Mundra, Panambur, Hazira, Port Blair, Alang, Visakhapatnam, Paradip, Ennore, Tuticorin and Nagapattinam are the other major ports in India. South Asian ports include Chittagong in Bangladesh, Colombo, Hambantota and Galle in Sri Lanka, and ports of Karachi, Sindh province and Gwadar, Balochistan province in Pakistan. Aden is a major port in Yemen and controls ships entering the Red Sea. Major African ports on the shores of the Indian Ocean include: Mombasa (Kenya), Dar es Salaam, Zanzibar (Tanzania), Durban, East London, Richard's Bay (South Africa), Beira (Mozambique), and Port Louis (Mauritius). Zanzibar is especially famous for its spice export. Other major ports in the Indian Ocean include Muscat (Oman), Yangon (Burma), Jakarta, Medan (Indonesia), Fremantle (port servicing Perth, Australia) and Dubai (UAE).

Chinese companies have made investments in several Indian Ocean ports, including Gwadar, Hambantota, Colombo and Sonadia. This has sparked a debate about the strategic implications of these investments.

Bordering Countries and Territories

Small islands dot the continental rims. Island nations within the ocean are Madagascar (the world's fourth largest island), Bahrain, Comoros, Maldives, Mauritius, Seychelles and Sri Lanka. The archipelago of Indonesia and the island nation of East Timor border the ocean on the east.

Heading roughly clockwise, the states and territories (in italics) with a coastline on the Indian Ocean (including the Red Sea and Persian Gulf) are:

Africa

·   South Africa

·   Mozambique

- Madagascar
- *French Southern and Antarctic Lands* (FRA)
- France (Mayotte and Réunion)
- Mauritius
- Comoros
- Tanzania
- Seychelles
- Kenya
- Somalia
- Djibouti
- Eritrea
- Sudan
- Egypt

Asia

- Egypt
- Israel
- Jordan
- Saudi Arabia
- Yemen
- Oman
- United Arab Emirates
- Qatar
- Bahrain
- Kuwait
- Iraq
- Iran
- Pakistan

- India
- Maldives
- *British Indian Ocean Territory* (UK)
- Sri Lanka
- Bangladesh
- Burma
- Thailand
- Malaysia
- Singapore
- Indonesia
- *Cocos (Keeling) Islands* (AUS)
- Timor-Leste

World Ocean

The World Ocean or global ocean (colloquially the sea or the ocean) is the interconnected system of Earth's oceanic waters, and comprises the bulk of the hydrosphere, covering almost 71% of Earth's surface, with a total volume of 1.332 billion cubic kilometers (351 quintillion US gallons).

Organization

The unity and continuity of the World Ocean, with relatively free interchange among its parts, is of fundamental importance to oceanography. It is divided into a number of principal oceanic areas that are delimited by the continents and various oceanographic features: these divisions are the Atlantic Ocean, Arctic Ocean (sometimes considered a sea of the Atlantic), Indian Ocean, Pacific Ocean, and Southern Ocean (often considered instead as just the southern portions of the Atlantic, Indian, and Pacific Oceans). In turn, oceanic waters are interspersed by many smaller seas, gulfs, and bays.

A global ocean has existed in one form or another on Earth for eons, and the notion dates back to classical antiquity in the form of Oceanus. The contemporary concept of the *World Ocean* was coined in the early 20th century by the Russian oceanographer Yuly Shokalsky to refer to the continuous ocean that covers and encircles most of Earth.

If viewed from the southern pole of Earth, the Atlantic, Indian, and Pacific Oceans can be seen as lobes extending northward from the Southern Ocean. Farther north, the Atlantic opens into the Arctic Ocean, which is connected to the Pacific by the Bering Strait, forming a continuous expanse of water.

- The Pacific Ocean, the largest of the oceans, also reaches northward from the Southern Ocean to the Arctic Ocean. It spans the gap between Australia and Asia, and the Americas. The Pacific Ocean meets the Atlantic Ocean south of South America at Cape Horn.

- The Atlantic Ocean, the second largest, extends from the Southern Ocean between the Americas, and Africa and Europe, to the Arctic Ocean. The Atlantic Ocean meets the Indian Ocean south of Africa at Cape Agulhas.

- The Indian Ocean, the third largest, extends northward from the Southern Ocean to India, between Africa and Australia. The Indian Ocean joins the Pacific Ocean to the east, near Australia.

- The Arctic Ocean is the smallest of the five. It joins the Atlantic Ocean near Greenland and Iceland and joins the Pacific Ocean at the Bering Strait. It overlies the North Pole, touching North America in the Western Hemisphere and Scandinavia and Siberia in the Eastern Hemisphere. The Arctic Ocean is partially covered in sea ice, the extent of which varies according to the season.

- The Southern Ocean is a proposed ocean surrounding Antarctica, dominated by the Antarctic Circumpolar Current, generally the ocean south of sixty degrees south latitude. The Southern Ocean is partially covered in sea ice, the extent of which varies according to the season. The Southern Ocean is the second smallest of the five named oceans.

Plate tectonics, post-glacial rebound and sea level rise continually change the coastline and structure of the world ocean.

Deepest Part of World Ocean

Mariana Trench

The Mariana Trench or Marianas Trench is the deepest part of the world's oceans. It is located in the western Pacific Ocean, to the east of the Mariana Islands. The trench is about 2,550 kilometres (1,580 mi) long with an average width of 69 kilometres (43 mi). It reaches a maximum-known depth of 10,994 m (± 40 m) or 6.831 mi (36,070 ± 131 ft) at a small slot-shaped valley in its floor known as the Challenger Deep, at its southern end, although some unrepeated measurements place the deepest portion at 11,034 metres (36,201 ft).

At the bottom of the trench the water column above exerts a pressure of 1,086 bars (15,750 psi) (over 1000 times the standard atmospheric pressure at sea level). At this pressure, the density of water is increased by 4.96%, making 95 litres of water under the pressure of the Challenger Deep contain the same mass as 100 litres at the surface. The temperature at the bottom is 1 to 4 °C.

The trench is not the part of the seafloor closest to the center of the Earth. This is because the Earth is not a perfect sphere; its radius is about 25 kilometres (16 mi) less at the poles than at the equator. As a result, parts of the Arctic Ocean seabed are at least 13 kilometres (8.1 mi) closer to the Earth's center than the Challenger Deep seafloor.

Xenophyophores have been found in the trench by Scripps Institution of Oceanography researchers at a record depth of 10.6 km (6.6 mi) below the sea surface. On 17 March 2013, researchers reported data that suggested microbial life forms thrive within the trench.

Names

The Mariana Trench is named for the nearby Mariana Islands (in turn named Las Marianas in honor of Spanish Queen Mariana of Austria, widow of Philip IV of Spain). The islands are part of the island arc that is formed on an over-riding plate, called the Mariana Plate (also named for the islands), on the western side of the trench.

Geology

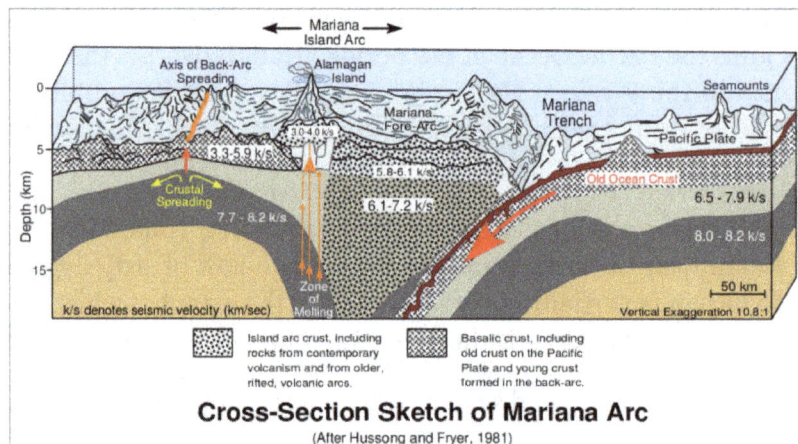

Cross-Section Sketch of Mariana Arc
(After Hussong and Fryer, 1981)

The Pacific plate is subducted beneath the Mariana Plate, creating the Mariana trench, and (further on) the arc of the Mariana islands, as water trapped in the plate is released and explodes upward to form island volcanoes.

The Mariana Trench is part of the Izu-Bonin-Mariana subduction system that forms the boundary between two tectonic plates. In this system, the western edge of one plate, the Pacific Plate, is subducted (i.e., thrust) beneath the smaller Mariana Plate that lies to the west. Crustal material at the western edge of the Pacific Plate is some of the oldest oceanic crust on earth (up to 170 million years old), and is therefore cooler and more dense; hence its great height difference relative to the higher-riding (and younger) Mariana Plate. The deepest area at the plate boundary is the Mariana Trench proper.

The movement of the Pacific and Mariana plates is also indirectly responsible for the formation of the Mariana Islands. These volcanic islands are caused by flux melting of the upper mantle due to release of water that is trapped in minerals of the subducted portion of the Pacific Plate.

Measurements

The trench was first sounded during the *Challenger* expedition in 1875, which recorded a depth of 4,475 fathoms (8.184 km; 26,850 ft). In 1877 a map was published called *Tiefenkarte des Grossen Ozeans* by Petermann, which showed a *Challenger Tief* at the location of that sounding. In 1899 USS Nero, a converted collier, recorded a depth of 5269 fathoms (9,636 m, 31,614 ft). *Challenger II* surveyed the trench using echo sounding, a much more precise and vastly easier way to measure depth than the sounding equipment and drag lines used in the original expedition. During this survey, the deepest part of the trench was recorded when the *Challenger II* measured a depth of 5,960 fathoms (10.90 km; 35,760 ft) at

11°19′N 142°15′E11.317°N 142.250°E, known as the Challenger Deep.

In 1957, the Soviet vessel *Vityaz* reported a depth of 11,034 m (36,201 ft), dubbed the *Mariana Hollow*.

In 1962, the surface ship M.V. *Spencer F. Baird* recorded a maximum depth of 10,915 m (35,840 ft), using precision depth gauges.

In 1984, the Japanese survey vessel *Takuyō* (拓洋) collected data from the Mariana Trench using a narrow, multi-beam echo sounder; it reported a maximum depth of 10,924 m (35,840 ft), also reported as 10,920 m (35,830 ft) ±10 m (33 ft),.

Remotely Operated Vehicle *KAIKO* reached the deepest area of Mariana trench and made the deepest diving record of 10,911 m (35,797 ft) on March 24, 1995.

During surveys carried out between 1997 and 2001, a spot was found along the Mariana Trench that had depth similar to that of the Challenger Deep, possibly even deeper. It was discovered while scientists from the Hawaii Institute of Geophysics and Planetology were completing a survey around Guam; they used a sonar mapping system towed behind the research ship to conduct the survey. This new spot was named the HMRG (Hawaii Mapping Research Group) Deep, after the group of scientists who discovered it.

On 1 June 2009 sonar mapping of the Challenger Deep by the Simrad EM120 sonar multibeam bathymetry system for deep water, i.e. depths of 300–11,000 m (980–36,000 ft) mapping aboard the *RV Kilo Moana* (mothership of the Nereus vehicle), has indicated a spot with a depth of 10,971 m (35,994 ft). The sonar system uses phase and amplitude bottom detection, with an accuracy of better than 0.2% of water depth across the entire swath (implying the depth figure is accurate to less than ± 22 metres).

In 2011, it was announced at the American Geophysical Union Fall Meeting that a US Navy hydrographic ship equipped with a multibeam echosounder conducted a survey which mapped the entire trench to 100 m (330 ft) resolution. The mapping revealed the existence of four rocky outcrops thought to be former seamounts.

The Mariana Trench is a site chosen by researchers at Washington University and the Woods Hole Oceanographic Institution in 2012 for a seismic survey to investigate the subsurface water cycle. Using both ocean-bottom seismometers and hydrophones the scientists are able to map structures as deep as 60 mi (97 km) beneath the surface.

Descents

Four descents have been achieved. The first was the manned descent by Swiss-designed, Italian-built, United States Navy-owned bathyscaphe *Trieste* which reached the bottom at 1:06 pm on 23 January 1960, with Don Walsh and Jacques Piccard on board. Iron shot was used for ballast, with gasoline for buoyancy. The onboard systems indicated a depth of 11,521 m (37,799 ft), but this was later revised to 10,916 m (35,814 ft). The depth was estimated from a conversion of pressure measured and calculations based on the water density from sea surface to seabed.

The bathyscaphe *Trieste* (designed by Auguste Piccard), the first manned vehicle to reach the bottom of the Marianas Trench.

This was followed by the unmanned ROVs *Kaikō* in 1996 and *Nereus* in 2009. The first three expeditions directly measured very similar depths of 10,902 to 10,916 m (35,768 to 35,814 ft).

The fourth was made by Canadian film director James Cameron in 2012. On 26 March, he reached the bottom of the Mariana Trench in the submersible vessel *Deepsea Challenger*.

In July 2015, members of the National Oceanic and Atmospheric Administration, Oregon State University, and the Coast Guard submerged a hydrophone into the deepest part of the Mariana Trench, the Challenger Deep. Never having deployed one past a mile, the titanium-shelled hydrophone was designed to withstand the immense pressure 7 miles under. Although researchers were unable to retrieve the hydrophone until November, the data capacity was full within the first 23 days. After months of analyzing the sounds, the experts were surprised to pick up natural *and* man-made sounds such as boats, earthquakes, a typhoon, and baleen whales. Due to the mission's success, the researchers plan to deploy a second hydrophone in 2017 for an extended period of time.

Planned Descents

As of February 2012, at least two other teams are planning piloted submarines to reach the bottom of the Mariana Trench. These include: Triton Submarines, a Florida-based company that designs and manufactures private submarines, for which a crew of three will take 120 minutes to reach the seabed; and DOER Marine, a marine technology company, based near San Francisco and set up in 1992, for which a crew of two or three will take 90 minutes to reach the seabed.

Life

The expedition conducted in 1960 claimed to have observed (with great surprise because of the high pressure) large creatures living at the bottom, such as a flatfish about 30 cm (1 ft) long, and shrimp. According to Piccard, "The bottom appeared light and clear, a waste of firm diatomaceous ooze". Many marine biologists are now skeptical of the supposed sighting of the flatfish, and it is suggested that the creature may instead have been a sea cucumber.

During the second expedition, the unmanned vehicle Kaikō collected mud samples from the seabed. Tiny organisms were found to be living in those samples.

In July 2011, a research expedition deployed untethered landers, called dropcams, equipped with digital video and lights to explore this region of the deep sea. Amongst many other living organisms, some gigantic single-celled amoebas with a size of more than 4 in (10 cm), belonging to the class of xenophyophores were observed. Xenophyophores are noteworthy for their size, their extreme abundance on the seafloor and their role as hosts for a variety of organisms.

In December 2014, a new species of snailfish was discovered at a depth of 8,145 m (26,722 ft), breaking the previous record for the deepest living fish seen on video. Several other new species were also filmed, including huge crustaceans known as supergiants.

Possible Nuclear Waste Disposal Site

Like other oceanic trenches, the Mariana Trench has been proposed as a site for nuclear waste disposal, in the hope that tectonic plate subduction occurring at the site might eventually push the nuclear waste deep into the Earth's mantle. However, ocean dumping of nuclear waste is prohibited by international law. Furthermore, plate subduction zones are associated with very large megathrust earthquakes, the effects of which are unpredictable and possibly adverse to the safety of long-term disposal. Also, disposal of nuclear wastes may cause havoc in the hadopelagic ecosystems.

Challenger Deep

The Challenger Deep is the deepest known point in the Earth's seabed hydrosphere, with a depth of 10,898 to 10,916 m (35,755 to 35,814 ft) by direct measurement from submersibles, and slightly more by sonar bathymetry. It is in the Pacific Ocean, at the southern end of the Mariana Trench near the Mariana Islands group. The Challenger Deep is a relatively small slot-shaped depression in the bottom of a considerably larger crescent-shaped oceanic trench, which itself is an unusually deep feature in the ocean floor. Its bottom is about 11 km (7 mi) long and 1.6 km (1 mi) wide, with gently sloping sides. The closest land to the Challenger Deep is Fais Island (one of the outer islands of Yap), 287 km (178 mi) southwest, and Guam, 304 km (189 mi) to the northeast. It is located in the ocean territory of the Federated States of Micronesia, 1.6 km (1 mi) from its border with ocean territory associated with Guam.

The depression is named after the British Royal Navy survey ship HMS *Challenger*, whose expedition of 1872–1876 made the first recordings of its depth. According to the August 2011 version of the GEBCO Gazetteer of Undersea Feature Names, the location and depth of the Challenger Deep are

11°22.4′N 142°35.5′E11.3733°N 142.5917°E and 10,920 m (35,827 ft) ±10 m (33 ft).

June 2009 sonar mapping of the Challenger Deep by the Simrad EM120 (sonar multibeam ba-thymetry system for 300–11,000 m deep water mapping) aboard the RV *Kilo Moana* indicated a depth of 10,971 metres (35,994 ft). The sonar system uses phase and amplitude bottom detection, with a precision of 0.2% to 0.5% of water depth; this is an error of about 22 to 55 m (72 to 180 ft) at this depth. Further soundings made by the US Center for Coastal & Ocean Mapping in October 2010 are in agreement with this figure, preliminarily placing the deepest part of the Challenger Deep at 10,994 m (36,070 ft), with an estimated vertical uncertainty of ±40 m (131 ft). A 2014 study concludes that with the best of 2010 multibeam echosounder technologies a depth uncer-tainty of ±25 m (82 ft) (95% confidence level) on 9 degrees of freedom and a positional uncertainty of ±20 to 25 m (66 to 82 ft) (2drms) remain and the location of the deepest depth recorded in the 2010 mapping is 10,984 m (36,037 ft) at

11°19′48″N 142°11′57″E11.329903°N 142.199305°E (

11°19′47.650″N 142°11′57.498″E11.32990278°N 142.19930500°E).

Only four descents have ever been achieved. The first descent by any vehicle was by the manned bathyscaphe *Trieste* in 1960. This was followed by the unmanned ROVs *Kaikō* in 1995 and *Nere-us* in 2009. In March 2012 a manned solo descent was made by the deep-submergence vehicle *Deepsea Challenger*. These expeditions measured very similar depths of 10,898 to 10,916 metres (35,755 to 35,814 ft).

History of Depth Mapping from the Surface

Over many years,the search for the point of maximum depth has involved many many vessels.

- The HMS *Challenger* expedition (December 1872 – May 1876) first sounded the depths now known as the Challenger Deep. This first sounding was made on 23 March 1875 at station 225. The reported depth was 4,475 fathoms (26,850 ft; 8,184 m) at 11°24′N 143°16′E11.400°N 143.267°E, based on two separate soundings.

- A 1912 book, *The Depths of the Ocean* by Sir John Murray, records the depth of the Challenger Deep as 31,614 ft (9,636 m), reporting the sounding taken by the converted navy collier USS *Nero* in 1899. Murray was one of the expedition scientists.

- In 1951, about 75 years after its original discovery, the entire Mariana Trench was surveyed by a second Royal Navy vessel, captained by George Stephen Ritchie (later Rear Admiral Ritchie); this vessel was also named HMS *Challenger*, after the original expedition ship. This survey recorded the deepest part of the trench using echo sounding, a more precise and easier way to measure depth than the sounding equipment and drag lines used in the original expedition. A depth of 5,960 fathoms (35,760 ft; 10,900 m) was measured at 11°19′N 142°15′E11.317°N 142.250°E.

Research vessel *Vityaz* in Kaliningrad "Museum of world ocean"

- The maximum surveyed depth of the Challenger Deep was reported in 1957 by the Soviet Research vessel *Vityaz* recording a spot 11,034 metres (36,201 ft) ±50 m (164 ft) deep at 11°20.9′N 142°11.5′E11.3483°N 142.1917°E. It was dubbed the *Mariana Hollow* and is listed in many reference sources, including the *Encyclopædia Britannica*, articles in *National Geographic* and on maps. The pressure at this depth is approximately 1,099 times that at the surface, or 111 MPa (16,099 psi).

· In 1959, the US Navy research vessel *RV Stranger* using bomb-sounding surveyed a maximum depth of 10,915 m (35,810 ft) ±10 m (33 ft) at 11°20.0′N 142°11.8′E11.3333°N 142.1967°E.

· In 1962, the US Navy research vessel *RV Spencer F. Baird* using a frequency-controlled depth recorder surveyed a maximum depth of 10,915 m (35,810 ft) ±10 m (33 ft) at 11°20.0′N 142°11.8′E11.3333°N 142.1967°E.

· In 1975 and 1980, the US Navy research vessel *RV Thomas Washington* using a precision depth recorder with satellite positioning surveyed a maximum depth of 10,915 m (35,810 ft) ±10 m (33 ft) at 11°20.0′N 142°11.8′E11.3333°N 142.1967°E.

· In 1984, the survey vessel *Takuyo* from the Hydrographic Department of Japan, used a narrow, multibeam echo sounder to take a measurement of 10,924 m (35,840 ft) ±10 m (33 ft) at 11°22.4′N 142°35.5′E11.3733°N 142.5917°E.

Deep Sea Research Vessel *RV Kairei*

· In 1998, a regional bathymetric survey of the Challenger Deep was conducted by the Deep Sea Research Vessel *RV Kairei*, from the Japan Agency for Marine-Earth Science and Technology, using a SeaBeam 2112 multibeam echosounder. The regional bathymetric map made from the data obtained in 1998 shows that the greatest depths in the eastern, central, and western depressions are 10,922 m (35,833 ft) ±74 m (243 ft), 10,898 m (35,755 ft) ±62 m (203 ft), and 10,908 m (35,787 ft) ±36 m (118 ft), respectively, making the eastern depression the deepest of the three.

· In 1999 and 2002, the *RV Kairei* revisited the Challenger Deep. The cross track survey in the 1999 *RV Kairei* cruise shows that the greatest depths in the eastern, central, and western depressions are 10,920 m (35,827 ft) ±10 m (33 ft), 10,894 m (35,741 ft) ±14 m (46 ft), and 10,907 m (35,784 ft) ±13 m (43 ft), respectively, which supports the results of the 1998 survey. The detailed grid survey in 2002 showed that the deepest site is located in the eastern part of the eastern depression around 11°22.260′N 142°35.589′E11.371000°N 142.593150°E, with a depth of 10,920 m (35,827 ft) ±5 m (16 ft), about 290 m (950 ft) southeast of the deepest site determined by the survey vessel *Takuyo* in 1984 and about 240 m (790 ft) east of the deepest place determined by the 1998 *RV Kairei* survey.

· On 1 June 2009, sonar mapping of the Challenger Deep by the Kongsberg Simrad EM 120 sonar multibeam bathymetry system for deep water (300 – 11,000 metres) mapping aboard the *RV Kilo Moana* (mothership of the *Nereus* underwater vehicle) indicated a depth of 10,971 m (35,994 ft). The sonar system uses phase and amplitude bottom detection, which is capable of an accuracy of 0.2% to 0.5% of water depth across the entire swath. In 2014 the multibeam bathymetry data of this sonar mapping have yet to be publicly released, so the data are not available for comparisons with other soundings.

· On 7 October 2010, further sonar mapping of the Challenger Deep area was conducted by the US Center for Coastal & Ocean Mapping/Joint Hydrographic Center (CCOM/ JHC) aboard the *USNS Sumner (T-AGS-61)*. The results were reported in December 2011 at the annual American Geophysical Union fall meeting. Using a Kongsberg Maritime EM 122 multibeam echosounder system coupled to positioning equipment that can determine latitude and longitude up to 50 cm (20 in) accuracy, from thousands of individual soundings around the deepest part the CCOM/JHC team preliminary determined that the Challenger Deep has a maximum depth of 10,994 m (36,070 ft) at 11°19′35″N 142°11′14″E11.326344°N 142.187248°E, with an estimated vertical uncertainty of ±40 m (131 ft) at 2 standard deviations (≈ 95.4%) confidence level. A secondary deep with a depth of 10,951 m (35,928 ft) was located at approximately 23.75 nmi (44.0 km) to the east at 11°22′11″N 142°35′19″E11.369639°N 142.588582°E in the Mariana Trench.

In 2014, a study was conducted regarding the determination of the depth and location of the Challenger Deep based on data collected previous to and during the 2010 sonar mapping of the Mariana Trench with a Kongsberg Maritime EM 122 multibeam echosounder system aboard the USNS Sumner (T-AGS-61). This study by James. V. Gardner et al. of the Center for Coastal & Ocean Mapping-Joint Hydrographic Center (CCOM/JHC), Chase Ocean Engineering Laboratory of the University of New Hampshire splits the measurement attempt history into three main groups: early single-beam echo sounders (1950s - 1970's), early multibeam echo sounders (1980s - 21st century), and modern (i.e., post-GPS, high-resolution) multibeam echo sounders. Taking uncertainties in depth measurements and position estimation into account the raw data of the 2010 bathymetry of the Challenger Deep vicinity consisting of 2,051,371 soundings from eight survey lines was analyzed. The study concludes that with the best of 2010 multibeam echosounder technologies after the analysis a depth uncertainty of ±25 m (82 ft) (95% confidence level) on 9 degrees of freedom and a positional uncertainty of ±20 to 25 m (66 to 82 ft) (2drms) remain and the location of the deepest depth recorded in the 2010 mapping is 10,984 m (36,037 ft) at

11°19′48″N 142°11′57″E11.329903°N 142.199305°E. The depth measurement uncertainty is a composite of measured uncertainties in the spatial variations in sound-speed through the water volume, the ray-tracing and bottom-detection algorithms of the multibeam system, the accuracies and calibration of the motion sensor and navigation systems, estimates of spherical spreading, attenuation throughout the water volume, and so forth.

The 2009 and 2010 maximal depths were not confirmed by the series of dives *Nereus* made to the bottom during an expedition in May–June 2009. The direct descent measurements by the four expeditions which have reported from the bottom, have fixed depths in a narrow range from 10,916 m (*Trieste*) to 10,911 m (*Kaikō*), to 10,902 m (*Nereus*) to 10,898 m (*Deepsea Challenger*) Although an attempt was made to correlate locations, it could not be absolutely certain that Nereus (or the other descents) reached exactly the same points found to be maximally deep by the sonar/echo sounders of previous mapping expeditions, even though one of these echo soundings was made by *Nereus* mothership.

Descents

Manned Descents

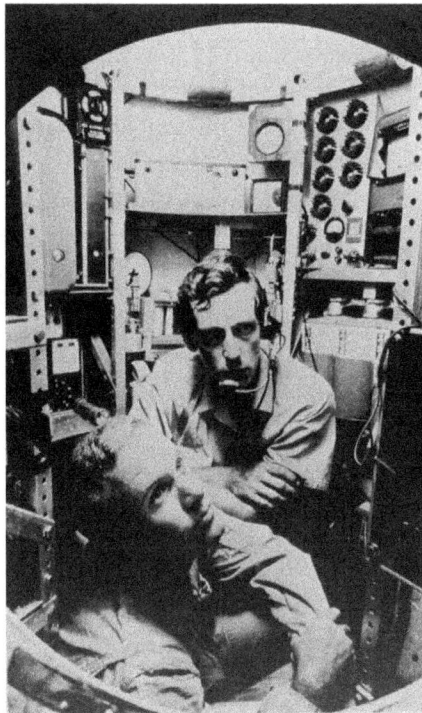

Lt. Don Walsh, USN (bottom) and Jacques Piccard (center) in the bathyscaphe *Trieste*.

Trieste

On 23 January 1960, the Swiss-designed *Trieste*, originally built in Italy and acquired by the U.S. Navy, descended to the ocean floor in the trench manned by Jacques Piccard (who co-designed the submersible along with his father, Auguste Piccard) and USN Lieutenant Don Walsh. Their crew compartment was inside a spherical pressure vessel, which was a heavy-duty replacement (of the Italian original) built by Krupp Steel Works of Essen, Germany. Their descent took almost five hours and the two men spent barely twenty minutes on the ocean floor before undertaking the three-hour-and-fifteen-minute ascent. Their early departure from the ocean floor was due to their concern over a crack in the outer window caused by the temperature differences during their descent. The measured depth at the bottom was measured with a manometer at 10,916 m (35,814 ft) ±5 m (16 ft).

Deepsea Challenger

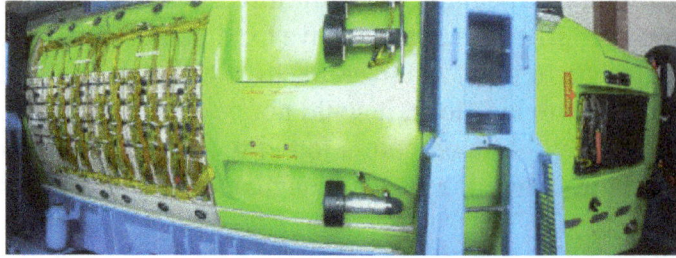

DSV Deepsea Challenger

On 26 March 2012 (local time), Canadian film director James Cameron made a solo manned descent in the DSV *Deepsea Challenger* to the bottom of the Challenger Deep. At approximately 05:15 ChST on 26 March (19:15 UTC on 25 March), the descent began. At 07:52 ChST (21:52 UTC), *Deepsea Challenger* arrived at the bottom. The descent lasted 2 hours and 36 minutes and the recorded depth was 10,898.4 metres (35,756 ft) when *Deepsea Challenger* touched down. Cameron had planned to spend about six hours near the ocean floor exploring but decided to start the ascent to the surface after only 2 hours and 34 minutes. The time on the bottom was shortened because a hydraulic fluid leak in the lines controlling the manipulator arm obscured the visibility out the only viewing port. It also caused the loss of the submersible's starboard thrusters. At around 12:00 ChST (02:00 UTC on 26 March), the Deepsea Challenge website says the sub resurfaced after a 90-minute ascent, although Paul Allen's tweets indicate the ascent took only about 67 minutes. During a post-dive press conference Cameron said: "I landed on a very soft, almost gelatinous flat plain. Once I got my bearings, I drove across it for quite a distance ... and finally worked my way up the slope." The whole time, Cameron said, he didn't see any fish, or any living creatures more than an inch (2.5 cm) long: "The only free swimmers I saw were small amphipods"—shrimplike bottom-feeders.

Planned Manned Descents

Several other manned expeditions are planned. These include:

- Triton Submarines, a Florida-based company that designs and manufactures private submarines, whose vehicle, Triton 36000/3, will carry a crew of three to the seabed in 120 minutes;

- Virgin Oceanic, sponsored by Richard Branson's Virgin Group, is developing a submersible designed by Graham Hawkes, DeepFlight Challenger, with which the solo pilot will take 140 minutes to reach the seabed;

- DOER Marine, a San Francisco Bay Area based marine technology company established in 1992, that is developing a vehicle, Deepsearch (and Ocean Explorer HOV Unlimited), with some support from Google's Eric Schmidt with which a crew of two or three will take 90 minutes to reach the seabed, as the program Deep Search.

Unmanned Descents

Kaikō

On 24 March 1995, the Japanese robotic deep-sea probe *Kaikō* broke the depth record for unmanned probes when it reached close to the surveyed bottom of the Challenger Deep. Created by the Japan Agency for Marine-Earth Science and Technology (JAMSTEC), it was one of the few unmanned deep-sea probes in operation that could dive deeper than 6,000 metres (20,000 ft). The manometer measured depth of 10,911 m (35,797 ft) ±3 m (10 ft) at

11°22.39′N 142°35.54′E11.37317°N 142.59233°E for the Challenger Deep is believed to be the most accurate measurement taken yet. *Kaikō* also collected sediment cores containing marine organisms from the bottom of the deep. *Kaikō* made many unmanned descents to the Mariana Trench during three expeditions in 1995, 1996 and 1998. The greatest depth measured by *Kaikō* in 1996 was 10,898 m (35,755 ft) at

11°22.10′N 142°25.85′E11.36833°N 142.43083°E and in 1998 10,907 m (35,784 ft) at

11°22.95′N 142°12.42′E11.38250°N 142.20700°E.

Abismo

On 3 June 2008, the Japanese robotic deep-sea probe *ABISMO* (Automatic Bottom Inspection and Sampling Mobile) reached the bottom of the Mariana Trench about 150 km (93 mi) east of the Challenger Deep and collected core samples of the deep sea sediment and water samples of the water column. Created by the Japan Agency for Marine-Earth Science and Technology (JAMSTEC), it is the only unmanned deep-sea probe in use that can dive deeper than 10,000 m (32,808 ft) after the loss of *Nereus*. During *ABISMO's* deepest Mariana Trench dive its manometer measured a depth of 10,258 m (33,655 ft) ±3 m (10 ft)

Nereus

HROV Nereus

On 31 May 2009 the United States sent the *Nereus* hybrid remotely operated vehicle (HROV) to the Challenger Deep. Nereus thus became the first vehicle to reach the Mariana Trench since 1998 and the deepest-diving vehicle then in operation. Project manager and developer Andy Bowen heralded the achievement as "the start of a new era in ocean exploration". *Nereus*, unlike *Kaikō*, did not need to be powered or controlled by a cable connected to a ship on the ocean surface.

Nereus spent over 10 hours at the bottom of the Challenger Deep and measured a depth of 10,902 m (35,768 ft) at

11°22.1′N 142°35.4′E11.3683°N 142.5900°E, while sending live video and data back to its mothership *RV Kilo Moana* at the surface and collecting geological and biological samples from the Challenger Deep bottom with its manipulator arm for further scientific analysis.

The *Nereus* was operated by the Woods Hole Oceanographic Institution. It was lost on May 10, 2014.

Lifeforms

The Summary Report of the HMS *Challenger* expedition lists radiolaria from the two dredged samples taken when the Challenger Deep was first discovered. These (Nassellaria and Spumellaria) were reported in the Report on Radiolaria (1887) written by Ernst Haeckel.

On their 1960 descent, the crew of the *Trieste* noted that the floor consisted of diatomaceous ooze and reported observing "some type of flatfish" lying on the seabed.

"... And as we were settling this final fathom, I saw a wonderful thing. Lying on the bottom just beneath us was some type of flatfish, resembling a sole, about 1 foot long and 6 inches across. Even as I saw him, his two round eyes on top of his head spied us — a monster of steel — invading his silent realm. Eyes? Why should he have eyes? Merely to see phosphorescence? The floodlight that bathed him was the first real light ever to enter this hadal realm. Here, in an instant, was the answer that biologists had asked for the decades. Could life exist in the greatest depths of the ocean? It could! And not only that, here apparently, was a true, bony teleost fish, not a primitive ray or elasmobranch. Yes, a highly evolved vertebrate, in time's arrow very close to man himself. Slowly, extremely slowly, this flatfish swam away. Moving along the bottom, partly in the ooze and partly in the water, he disappeared into his night. Slowly too — perhaps everything is slow at the bottom of the sea — Walsh and I shook hands.

Many marine biologists are now skeptical of this supposed sighting, and it is suggested that the creature may instead have been a sea cucumber. The video camera on board the *Kaiko* probe spotted a sea cucumber, a scale worm and a shrimp at the bottom. At the bottom of the Challenger deep, the *Nereus* probe spotted one polychaete worm (a multi-legged predator) about an inch long.

An analysis of the sediment samples collected by *Kaiko* found large numbers of simple organisms at 10,900 m (35,800 ft). While similar lifeforms have been known to exist in shallower ocean trenches (> 7,000 m) and on the abyssal plain, the lifeforms discovered in the Challenger Deep possibly represent taxa distinct from those in shallower ecosystems.

Most of the organisms collected were simple, soft-shelled foraminifera (432 species according to National Geographic), with four of the others representing species of the complex, multi-chambered genera *Leptohalysis* and *Reophax*. Eighty-five percent of the specimens were organic, soft-shelled allogromiids, which is unusual compared to samples of sediment-dwelling organisms from other deep-sea environments, where the percentage of organic-walled foraminifera ranges from 5% to 20%. As small organisms with hard, calcareous shells have trouble growing at extreme depths because of the high solubility of calcium carbonate in the pressurized water, scientists theorize that the preponderance of soft-shelled organisms in the Challenger Deep may have resulted from the typical biosphere present when the Challenger Deep was shallower than it is now. Over the course of six to nine million years, as the Challenger Deep grew to its present depth, many of the species present in the sediment died out or were unable to adapt to the increasing water pressure and changing environment.} The species that survived the change in depth were the ancestors of the Challenger Deep's current denizens.

On 17 March 2013, researchers reported data that suggested microbial life forms thrive in the Challenger Deep. Other researchers reported related studies that microbes thrive inside rocks up to 1900 feet below the sea floor under 8500 feet of ocean off the coast of the northwestern United States. According to one of the researchers, "You can find microbes everywhere — they're extremely adaptable to conditions, and survive wherever they are."

Borders of the Oceans

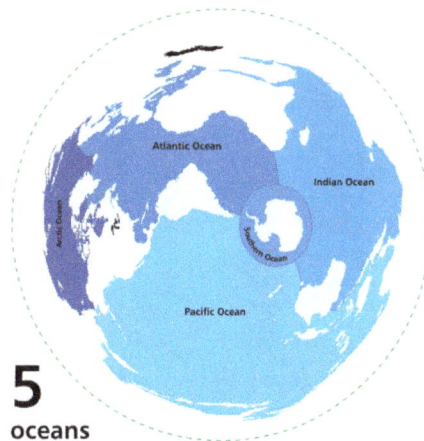

5 oceans

Maps exhibiting the world's oceanic waters. A continuous body of water encircling the Earth, the world (global) ocean is divided into a number of principal areas. Five oceanic divisions are usually recognized: Pacific, Atlantic, Indian, Arctic, and Southern; the last two listed are sometimes consolidated into the first three.

The borders of the oceans are the limits of the Earth's oceanic waters. The definition and number of oceans can vary depending on the adopted criteria.

Overview

Though generally described as several separate oceans, the world's oceanic waters constitute one global, interconnected body of salt water sometimes referred to as the World Ocean or global ocean. This concept of a continuous body of water with relatively free interchange among its parts is of fundamental importance to oceanography.

The major oceanic divisions are defined in part by the continents, various archipelagos, and other criteria. The principal divisions (in descending order of area) are the: Pacific Ocean, Atlantic Ocean, Indian Ocean, Arctic Ocean, and Southern (Antarctic) Ocean. Smaller regions of the oceans are called seas, gulfs, bays, straits, and other names.

Geologically, an ocean is an area of oceanic crust covered by water. Oceanic crust is the thin layer of solidified volcanic basalt that covers the Earth's mantle. Continental crust is thicker but less dense. From this perspective, the Earth has three oceans: the World Ocean, the Caspian Sea, and Black Sea. The latter two were formed by the collision of Cimmeria with Laurasia. The Mediterranean Sea is at times a discrete ocean, because tectonic plate movement has repeatedly broken its connection to the World Ocean through the Strait of Gibraltar. The Black Sea is connected to the Mediterranean through the Bosporus, but the Bosporus is a natural canal cut through continental rock some 7,000 years ago, rather than a piece of oceanic sea floor like the Strait of Gibraltar.

Despite their names, some smaller landlocked "seas" are *not* connected with the World Ocean, such as the Caspian Sea and numerous salt lakes such as the Aral Sea.

A complete hierarchy showing which seas belong to which oceans, according to the International Hydrographic Organization (IHO) and for the whole planet, is available at the European Marine Gazetteer website.

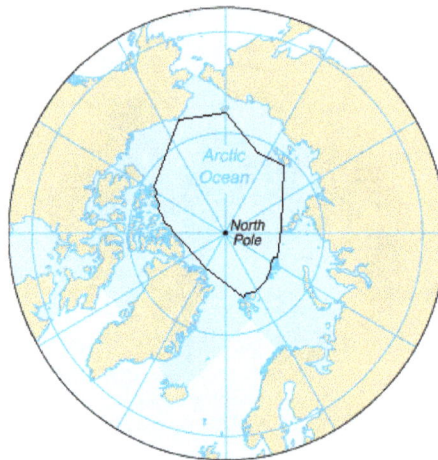

The borders of the Arctic Ocean, according to the CIA *The World Factbook* (blue area), and as defined by the IHO (black outline - excluding marginal waterbodies).

Arctic Ocean

The Arctic Ocean covers much of the Arctic and washes upon northern North America and Eurasia and is sometimes considered a sea or estuary of the Atlantic.

The International Hydrographic Organization (IHO) defines the limits of the Arctic Ocean (excluding the seas it contains) as follows:

> *Between Greenland and West Spitzbergen* — The Northern limit of Greenland Sea.

> *Between West Spitzbergen and North East Land* — the parallel of lat. 80°N.

> *From Cape Leigh Smith to Cape Kohlsaat* — the Northern limit of Barentsz Sea .

> *From Cape Kohlsaat to Cape Molotov* — the Northern limit of Kara Sea.

> *From Cape Molotov to the Northern extremity of Kotelni Island* — the Northern limit of Laptev Sea.

From the Northern extremity of Kotelni Island to the Northern point of Wrangel Island — the Northern limit of East Siberian Sea.

From the Northern point of Wrangel Island to Point Barrow — the Northern limit of Chuckchi Sea.

From Point Barrow to Cape Land's End on Prince Patrick Island — the Northern limit of Beaufort Sea,[Ar 7] through the Northwest coast of Prince Patrick Island to Cape Leopold M'Clintock, thence to Cape Murray (Brook Island) and along the Northwest coast to the extreme Northerly point; to Cape Mackay (Borden Island); through the Northwesterly coast of Borden Island to Cape Malloch, to Cape Isachsen (Ellef Ringnes Island); to the Northwest point of Meighen Island to Cape Stallworthy (Axel Heiberg Island) to Cape Colgate the extreme West point of Ellesmere Island; through the North shore of Ellesmere Island to Cape Columbia thence a line to Cape Morris Jesup (Greenland).

Note that these definitions exclude any marginal waterbodies that are separately defined by the IHO (such as the Kara Sea and East Siberian Sea), though these are usually considered to be part of the Arctic Ocean.

The CIA defines the limits of the Arctic Ocean differently, as depicted in the map comparing its definition to the IHO's definition.

Atlantic Ocean

The Atlantic Ocean, according to the CIA *The World Factbook* (blue area), and as defined by the IHO (black outline - excluding marginal waterbodies).

The Atlantic Ocean separates the Americas from Eurasia and Africa. It may be further subdivided by the equator into northern and southern portions.

North Atlantic

The 3rd edition, currently in force, of the International Hydrographic Organization's (IHO) *Limits of Oceans and Seas* defines the limits of the North Atlantic Ocean (excluding the seas it contains) as follows:

> *On the West.* The Eastern limits of the Caribbean Sea, the Southeastern limits of the Gulf of Mexico from the North coast of Cuba to Key West, the Southwestern limit of the Bay of Fundy and the Southeastern and Northeastern limits of the Gulf of St. Lawrence.
>
> *On the North.* The Southern limit of Davis Strait from the coast of Labrador to Greenland and the Southwestern limit of the Greenland Sea and Norwegian Sea from Greenland to the Shetland Islands.
>
> *On the East.* The Northwestern limit of the North Sea, the Northern and Western limits of the Scottish Seas, the Southern limit of the Irish Sea, the Western limits of the Bristol and English Channels, of the Bay of Biscay and of the Mediterranean Sea.
>
> *On the South.* The equator, from the coast of Brazil to the Southwestern limit of the Gulf of Guinea.

South Atlantic

The 3rd edition (currently in force) of the International Hydrographic Organization's (IHO) *Limits of Oceans and Seas* defines the limits of the South Atlantic Ocean (excluding the seas it contains) as follows:

> *On the Southwest.* The meridian of Cape Horn, Chile (67°16'W) from Tierra del Fuego to the Antarctic Continent; a line from Cape Virgins (52°21'S 68°21'W52.350°S 68.350°W) to Cape Espiritu Santo, Tierra del Fuego, the Eastern entrance to Magellan Strait, Chile
>
> *On the West.* The limit of the Rio de La Plata.
>
> *On the North.* The Southern limit of the North Atlantic Ocean.
>
> *On the Northeast.* The limit of the Gulf of Guinea.
>
> *On the Southeast.* From Cape Agulhas along the meridian of 20° East to the Antarctic continent.
>
> *On the South.* The Antarctic Continent.

Note that these definitions exclude any marginal waterbodies that are separately defined by the IHO (such as the Bay of Biscay and Gulf of Guinea), though these are usually considered to be part of the Atlantic Ocean.

In its 2002 draft, the IHO redefined the Atlantic Ocean, moving its southern limit to 60°S, with the waters south of that line identified as the Southern Ocean. This new definition has not yet been ratified (and, in addition, a reservation was lodged in 2003 by Australia.) While the name "Southern Ocean" is frequently used, some geographic authorities such as the 10th edition of the World Atlas from the U.S. National Geographic Society generally show the Atlantic, Indian, and Pacific Oceans continuing to Antarctica. If and when adopted, the 2002 definition would be published in the 4th edition of *Limits of Oceans and Seas*, re-instituting the 2nd edition's "Southern Ocean", omitted from the 3rd edition.

Indian Ocean

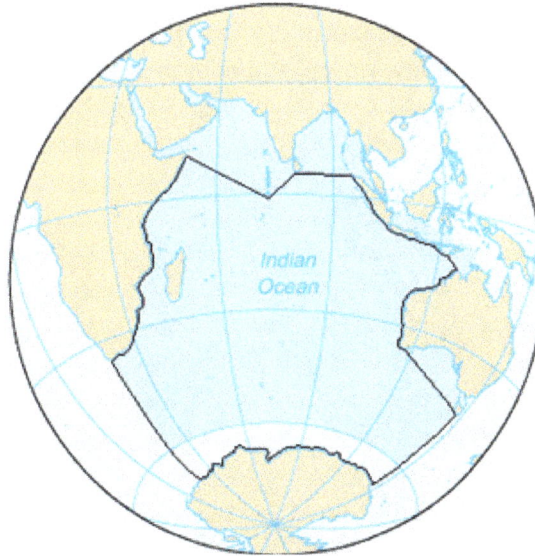

The Indian Ocean, according to the CIA The World Factbook (blue area), and as defined by the IHO (black outline - excluding marginal waterbodies).

The Indian Ocean washes upon southern Asia and separates Africa and Australia.

The 3rd edition, currently in force, of the International Hydrographic Organization's (IHO) *Limits of Oceans and Seas* defines the limits of the Indian Ocean (excluding the seas it contains) as follows:

> *On the North.* The Southern limits of the Arabian Sea and the Lakshadweep Sea, the Southern limit of the Bay of Bengal, the Southern limits of the East Indian Archipelago, and the Southern limit of the Great Australian Bight.

> *On the West.* From Cape Agulhas in 20° long. East, Southward along this meridian to the Antarctic Continent.

> *On the East.* From South East Cape, the Southern point of Tasmania down the meridian 146°55'E to the Antarctic Continent.

> *On the South.* The Antarctic Continent.

Note that this definition excludes any marginal waterbodies that are separately defined by the IHO (such as the Bay of Bengal and Arabian Sea), though these are usually considered to be part of the Indian Ocean.

In its 2002 draft, the IHO redefined the Indian Ocean, moving its southern limit to 60°S, with the waters south of that line identified as the Southern Ocean. This new definition has not yet been ratified (and, in addition, a reservation was lodged in 2003 by Australia.) While the name "Southern Ocean" is frequently used, some geographic authorities such as the 10th edition of the World Atlas from the U.S. National Geographic Society generally show the Atlantic, Indian, and Pacific Oceans continuing to Antarctica. If and when adopted, the 2002 definition would be published in the 4th edition of *Limits of Oceans and Seas*, re-instituting the 2nd edition's "Southern Ocean", omitted from the 3rd edition.

The boundary of the Indian Ocean is a constitutional issue for Australia. The Imperial South Australia Colonisation Act, 1834, which established and defined the Colony of South Australia defined South Australia's southern limit as being the "Southern Ocean." This definition was carried through to Australian constitutional law upon the Federation of Australia in 1901.

Pacific Ocean

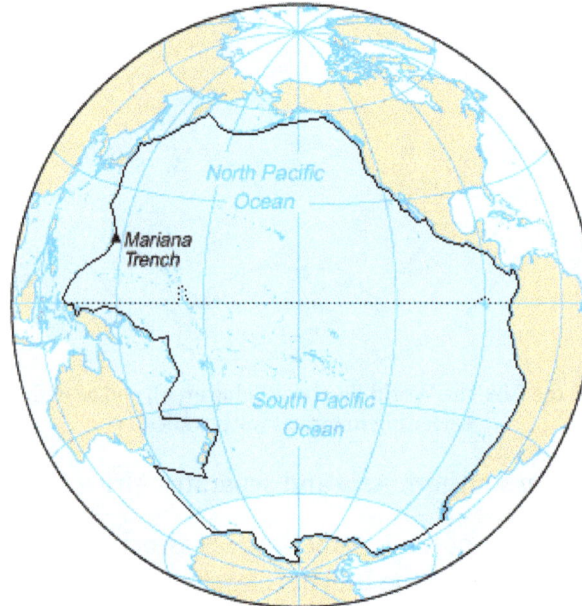

The Pacific Ocean according to the CIA *The World Factbook* (blue area), and as defined by the IHO (black outline — excluding marginal waterbodies)

The Pacific is the ocean that separates Asia and Australia from the Americas. It may be further subdivided by the equator into northern and southern portions.

North Pacific

The 3rd edition, currently in force, of the International Hydrographic Organization's (IHO) *Limits of Oceans and Seas* defines the limits of the North Pacific Ocean (excluding the seas it contains) as follows:

> *On the Southwest.* The Northeastern limit of the East Indian Archipelago from the Equator to Luzon Island.

> *On the West and Northwest.* The Eastern limits of the Philippine Sea[P 1] and Japan Sea and the Southeastern limit of the Sea of Okhotsk.

> *On the North.* The Southern limits of the Bering Sea[P 4] and the Gulf of Alaska.

> *On the East.* The Western limit of Coastal waters of Southeast Alaska and Br. Columbia, and the Southern limit of the Gulf of California.

> *On the South.* The Equator, but excluding those islands of the Gilbert and Galàpagos Groups which lie to the Northward thereof.

South Pacific

The 3rd edition, currently in force, of the International Hydrographic Organization's (IHO) *Limits of Oceans and Seas* defines the limits of the South Pacific Ocean (excluding the seas it contains) as follows:

> *On the West.* From Southeast Cape, the Southern point of Tasmania, down the meridian of 146°55'E to the Antarctic continent.

> *On the Southwest and Northwest.* The Southern, Eastern and Northeastern limits of the Tasman Sea, the Southeastern and Northeastern limits of the Coral Sea, the Southern, Eastern and Northern limits of the Solomon and Bismark seas, and the Northeastern limit of the East Indian Archipelago from New Guinea to the Equator.

> *On the North.* The Equator, but including those islands of the Gilbert and Galàpagos Groups which lie to the Northward thereof.

> *On the East.* The meridian of Cape Horn (67°16'W) from Tierra del Fuego to the Antarctic continent; a line from Cape Virgins (52°21'S 68°21'W52.350°S 68.350°W) to Cape Espititu Santo, Tierra del Fuego, the Eastern entrance to Magellan Strait. (These limits have not yet been officially accepted by Argentina and Chile.)

> *On the South.* The Antarctic continent.

Note that these definitions exclude any marginal waterbodies that are separately defined by the IHO (such as the Gulf of Alaska and Coral Sea), though these are usually considered to be part of the Pacific Ocean.

In its 2002 draft, the IHO redefined the Pacific Ocean, moving its southern limit to 60°S, with the waters south of that line identified as the Southern Ocean. This new definition has not yet been ratified (and, in addition, a reservation was lodged in 2003 by Australia.) While the name "Southern Ocean" is frequently used, some geographic authorities such as the 10th edition of the World Atlas from the U.S. National Geographic Society generally show the Atlantic, Indian, and Pacific Oceans continuing to Antarctica. If and when adopted, the 2002 definition would be published in the 4th edition of *Limits of Oceans and Seas*, re-instituting the 2nd edition's "Southern Ocean", omitted from the 3rd edition.

Southern or Antarctic Ocean

The Southern Ocean contains the waters that surround Antarctica and sometimes is considered an extension of Pacific, Atlantic and Indian Oceans.

In 1928, the first edition of the International Hydrographic Organization's (IHO) *Limits of Oceans and Seas* publication included the Southern Ocean around Antarctica. The Southern Ocean was delineated by land-based limits - the continent of Antarctica to the south, and the continents of South America, Africa, and Australia plus Broughton Island, New Zealand in the north. The detailed land-limits used were Cape Horn in South America, Cape Agulhas in Africa, the southern coast of Australia from Cape Leeuwin, Western Australia, to South East Cape, Tasmania, via the western edge of the water body of Bass Strait, and then Broughton Island before returning to Cape Horn.

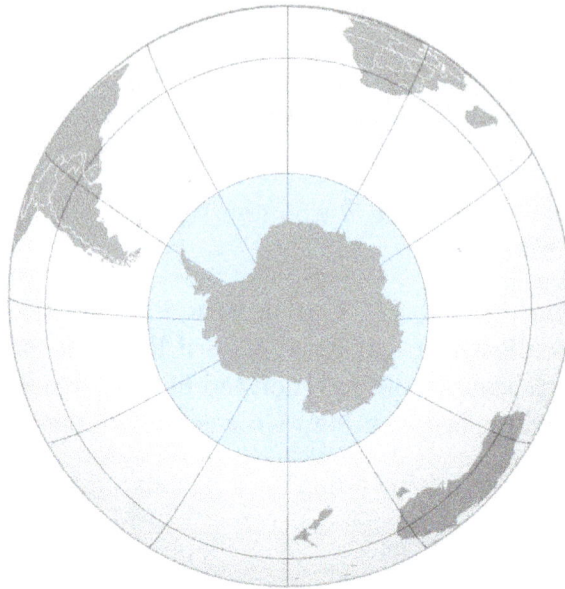

The Southern Ocean according to the CIA *The World Factbook*

The northern limits of the Southern Ocean were moved southwards in the IHO's 1937 second edition of the *Limits of Oceans and Seas*. The Southern Ocean then extended from Antarctica northwards to latitude 40° south between Cape Agulhas in Africa (long. 20° east) and Cape Leeuwin in Western Australia (long. 115° east), and extended to latitude 55° south between Auckland Island of New Zealand (long. 165° or 166° east) and Cape Horn in South America (long. 67° west).

The Southern Ocean did not appear in the 1953 third edition because "*...the northern limits ... are difficult to lay down owing to their seasonal change ... Hydrographic Offices who issue separate publications dealing with this area are therefore left to decide their own northern limits. (Great Britain uses the Latitude of 55° South)*". Instead, in the IHO 1953 publication, the Atlantic, Indian and Pacific Oceans were extended southward, the Indian and Pacific Oceans (which had not previously touched pre 1953, as per the first and second editions) now abutted at the meridian of South East Cape, and the southern limits of the Great Australian Bight and the Tasman Sea were moved northwards.

The IHO readdressed the question of the Southern Ocean in a survey in 2000. Of its 68 member nations, 28 responded, and all responding members except Argentina agreed to redefine the ocean, reflecting the importance placed by oceanographers on ocean currents. The proposal for the name *Southern Ocean* won 18 votes, beating the alternative *Antarctic Ocean*. Half of the votes supported a definition of the ocean's northern limit at 60°S (with no land interruptions at this latitude), with the other 14 votes cast for other definitions, mostly 50°S, but a few for as far north as 35°S.

The 4th edition of *Limits of Oceans and Seas* has yet to be published due to 'areas of concern' by several countries relating to various naming issues around the world. The IHB circulated a new draft of the 4th edition of the publication in August 2002, however there were still various changes, 60 seas were added or renamed from the 3rd edition, and even the name of the publication was changed. A reservation had also been lodged by Australia regarding the Southern Ocean limits. Effectively, the 3rd edition (which did not delineate the Southern Ocean leaving delineation to local hydrographic offices) has yet to be superseded and IHO documents declare that it remains "currently in force."

Despite this, the 4th edition definition has *de facto* usage by many organisations, scientists and nations - even at times by IHO committees. Some nations' hydrographic offices have defined their own boundaries; the United Kingdom used the 55°S parallel for example.

Other sources, such as the National Geographic Society, show the Atlantic, Pacific and Indian Oceans as extending to Antarctica, although articles on the National Geographic web site have begun to reference the Southern Ocean.

In Australia, cartographic authorities defined the Southern Ocean as including the entire body of water between Antarctica and the south coasts of Australia and New Zealand. This delineation is basically the same as the original (first) edition of the IHO publication and effectively the same as the second edition. In the second edition, the Great Australian Bight was defined as the only geographical entity between the Australian coast and the Southern Ocean. Coastal maps of Tasmania and South Australia label the sea areas as *Southern Ocean*, while Cape Leeuwin in Western Australia is described as the point where the Indian and Southern Oceans meet.

References

- Charette, Matthew; Smith, Walter H. F. (2010). "The volume of Earth's ocean". Oceanography. 23 (2): 112–114. doi:10.5670/oceanog.2010.51. Retrieved 27 September 2012.

- Tomczak, Matthias; Godfrey, J. Stuart (2003). Regional Oceanography: an Introduction (2 ed.). Delhi: Daya Publishing House. ISBN 81-7035-306-8.

- "Scientists map Mariana Trench, deepest known section of ocean in the world". The Telegraph. Telegraph Media Group. 7 December 2011. Retrieved 23 March 2012.

- Platt, Jane; Bell, Brian (2014-04-03). "NASA Space Assets Detect Ocean inside Saturn Moon". NASA. Retrieved 2014-04-03.

- Iess, L.; Stevenson, D.J.; Parisi, M.; Hemingway, D.; et al. (4 April 2014). "The Gravity Field and Interior Structure of Enceladus". Science. 344 (6179): 78–80. doi:10.1126/science.1250551. Retrieved 3 April 2014.

- Dyches, Preston; Chou, Felcia (7 April 2015). "The Solar System and Beyond is Awash in Water". NASA. Retrieved 8 April 2015.

- Wiktorowicz, Sloane J.; Ingersoll, Andrew P. (2007). "Liquid water oceans in ice giants". Icarus. 186 (2): 436–447. arXiv:astro-ph/0609723. Bibcode:2007Icar..186..436W. doi:10.1016/j.icarus.2006.09.003. ISSN 0019-1035.

- Silvera, Isaac (2010). "Diamond: Molten under pressure". Nature Physics. 6 (1): 9–10. Bibcode:2010NatPh...6....9S. doi:10.1038/nphys1491. ISSN 1745-2473.

- Clavin, Whitney (May 1, 2014). "Ganymede May Harbor 'Club Sandwich' of Oceans and Ice". NASA. Jet Propulsion Laboratory. Retrieved 2014-05-01.

- McKinnon, William B.; Kirk, Randolph L. (2007). "Triton". In Lucy Ann Adams McFadden; Lucy-Ann Adams; Paul Robert Weissman; Torrence V. Johnson. Encyclopedia of the Solar System (2nd ed.). Amsterdam; Boston: Academic Press. pp. 483–502. ISBN 978-0-12-088589-3.

- "The Inside Story". pluto.jhuapl.edu — NASA New Horizons mission site. Johns Hopkins University Applied Physics Laboratory. 2013. Retrieved 2 August 2013.

- Aguilar, David A. (2009-12-16). "Astronomers Find Super-Earth Using Amateur, Off-the-Shelf Technology". Harvard-Smithsonian Center for Astrophysics. Retrieved January 23, 2010.

- McKinnon, William B.; Kirk, Randolph L. (2007). "Triton". In Lucy Ann Adams McFadden; Lucy-Ann Adams; Paul Robert Weissman; Torrence V. Johnson. Encyclopedia of the Solar System (2nd ed.). Amsterdam; Boston: Academic Press. p. 485. ISBN 978-0-12-088589-3.

- "Study of the ice shells and possible subsurface oceans of the Galilean satellites using laser altimeters on board the Europa and Ganymede orbiters JEO and JGO" (PDF). Retrieved 2011-10-14.

- Tomczak, Matthias; Godfrey, J. Stuart (2003). Regional Oceanography: an Introduction (2nd ed.). Delhi: Daya Publishing House. ISBN 81-7035-306-8.

- Wright, John W., ed. (2006). The New York Times Almanac (2007 ed.). New York, New York: Penguin Books. p. 455. ISBN 0-14-303820-6.

- "Arctic Ocean Fast Facts". wwf.pandora.org (World Wildlife Foundation). Archived from the original on 29 October 2010. Retrieved 2010-10-28.

- "The Mariana Trench – Oceanography". www.marianatrench.com. 2003-04-04. Archived from the original on 7 December 2006. Retrieved 2006-12-02.

Various Aspects of Ocean

Ocean can be best understood in confluence with the major topics listed in the following chapter. Oceans cover 71 % of the surface of earth. Some of the aspects are ocean current, ocean heat content and plate tectonics. This chapter helps the reader to broaden the existing knowledge on oceans.

Ocean Current

An ocean current is a continuous, directed movement of seawater generated by forces acting upon this mean flow, such as breaking waves, wind, the Coriolis effect, cabbeling, temperature and salinity differences, while tides are caused by the gravitational pull of the Sun and Moon. Depth contours, shoreline configurations, and interactions with other currents influence a current's direction and strength.

Ocean currents flow for great distances, and together, create the global conveyor belt which plays a dominant role in determining the climate of many of the Earth's regions. More specifically, ocean currents influence the temperature of the regions through which they travel. For example, warm currents traveling along more temperate coasts increase the temperature of the area by warming the sea breezes that blow over them. Perhaps the most striking example is the Gulf Stream, which makes northwest Europe much more temperate than any other region at the same latitude. Another example is Lima, Peru where the climate is cooler (sub-tropical) than the tropical latitudes in which the area is located, due to the effect of the Humboldt Current.

Function

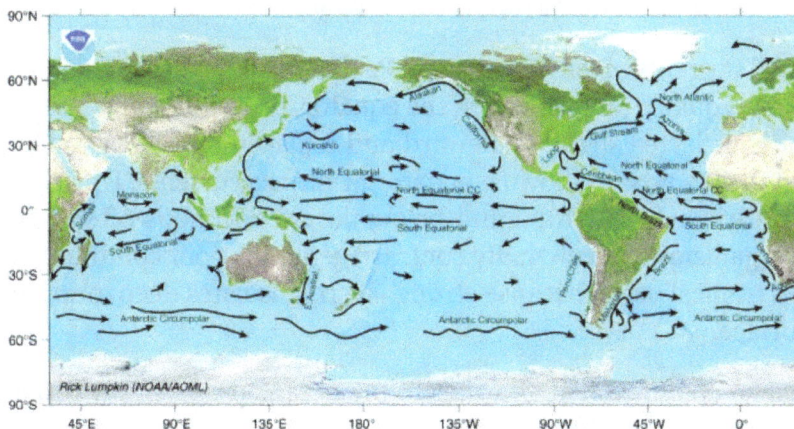

Major ocean surface currents (Source: NOAA)

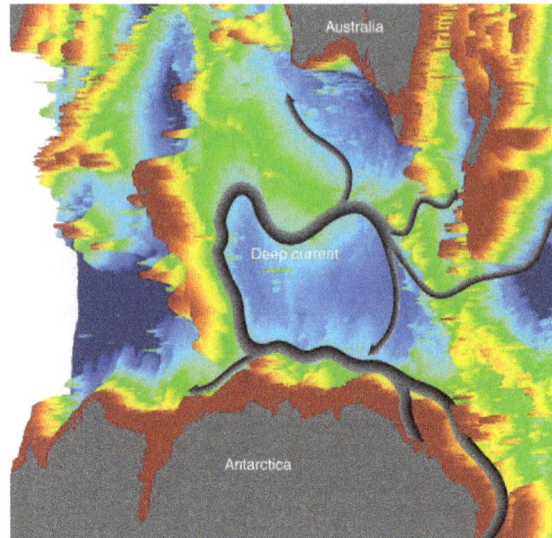

The bathymetry of the Kerguelen Plateau in the Southern Ocean governs the course of the new current part of the global network of ocean currents (Source:CSIRO)

Surface oceanic currents are sometimes wind driven and develop their typical clockwise spirals in the northern hemisphere and counter-clockwise rotation in the southern hemisphere because of imposed wind stresses. In wind driven currents, the Ekman spiral effect results in the currents flowing at an angle to the driving winds. The areas of surface ocean currents move somewhat with the seasons; this is most notable in equatorial currents.

Deep ocean basins generally have a non-symmetric surface current, in that the eastern equatorward-flowing branch is broad and diffuse whereas the western poleward flowing branch is very narrow. These western boundary currents (of which the Gulf Stream is an example) are a consequence of the rotation of the Earth.

Deep ocean currents are driven by density and temperature gradients. Thermohaline circulation is also known as the ocean's conveyor belt (which refers to deep ocean density driven ocean basin currents). These currents, called submarine rivers, flow under the surface of the ocean and are hidden from immediate detection. Where significant vertical movement of ocean currents is observed, this is known as upwelling and downwelling. Deep ocean currents are currently being researched using a fleet of underwater robots called Argo.

The South Equatorial Currents of the Atlantic and Pacific straddle the equator. Though the Coriolis effect is weak near the equator (and absent at the equator), water moving in the currents on either side of the equator is deflected slightly poleward and replaced by deeper water. Thus, equatorial upwelling occurs in these westward flowing equatorial surface currents. Upwelling is an important process because this water from within and below the pycnocline is often rich in the nutrients needed by marine organisms for growth. By contrast, generally poor conditions for growth prevail in most of the open tropical ocean because strong layering isolates deep, nutrient rich water from the sunlit ocean surface.

Surface currents make up only 8% of all water in the ocean, are generally restricted to the upper 400 m (1,300 ft) of ocean water, and are separated from lower regions by varying temperatures and salinity which affect the density of the water, which in turn, defines each oceanic region. Because the movement of deep water in ocean basins is caused by density driven forces and gravity, deep waters sink into deep ocean basins at high latitudes where the temperatures are cold enough to cause the density to increase.

Ocean currents are measured in sverdrup (sv), where 1 sv is equivalent to a volume flow rate of 1,000,000 m³ (35,000,000 cu ft) per second.

Surface currents are found on the surface of an ocean, and are driven by large scale wind currents. They are directly affected by the wind—the Coriolis effect plays a role in their behaviors.

Thermohaline Circulation

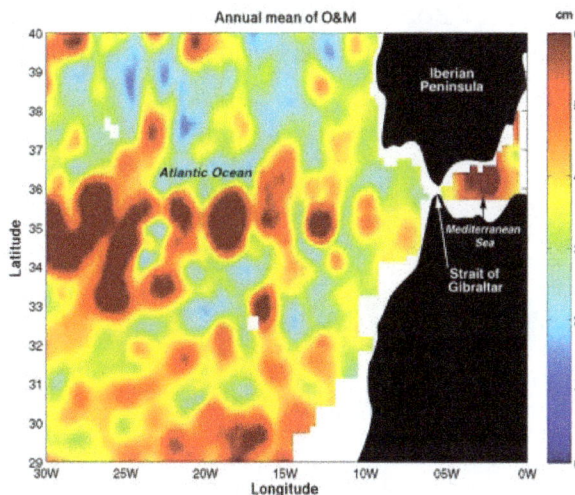

Coupling data collected by NASA/JPL by several different satellite-borne sensors, researchers have been able to "break through" the ocean's surface to detect "Meddies" -- super-salty warm-water eddies that originate in the Mediterranean Sea and then sink more than a half-mile underwater in the Atlantic Ocean. The Meddies are shown in red in this scientific figure.

Horizontal and vertical currents also exist below the pycnocline in the ocean's deeper waters. The movement of water due to differences in density as a function of water temperature and salinity is called thermohaline circulation. Ripple marks in sediments, scour lines, and the erosion of rocky outcrops on deep-ocean floors are evidence that relatively strong, localized bottom currents exist. Some of these currents may move as rapidly as 60 centimeters (24 inches) per second.

These currents are strongly influenced by bottom topography, since dense, bottom water must forcefully flow around seafloor projections. Thus, they are sometimes called contour currents. Bottom currents generally move equator-ward at or near the western boundaries of ocean basins (below the western boundary surface currents). The deep-water masses are not capable of moving water at speeds comparable to that of wind-driven surface currents. Water in some of these currents may move only 1 to 2 meters per day. Even at that slow speed, the Coriolis effect modifies their pattern of flow.

Downwelling of Deep Water in Polar Regions

Antarctic Bottom Water is the most distinctive of the deep-water masses. It is characterized by a salinity of 34.65‰, a temperature of -0.5 °C (30 °F), and a density of 1.0279 grams per cubic centimeter. This water is noted for its extreme density (the densest in the world ocean), for the great amount of it produced near Antarctic coasts, and for its ability to migrate north along the seafloor. Most Antarctic Bottom Water forms near the Antarctic coast south of South America during winter. Salt is concentrated in pockets between crystals of pure water and then squeezed out of the freezing mass to form a frigid brine. Between 20 million and 50 million cubic meters of this brine form every second. The water's great density causes it to sink toward the continental shelf, where it mixes with nearly equal parts of water from the southern Antarctic Circumpolar Current. The mixture settles along the edge of Antarctica's continental shelf, descends along the slope, and spreads along the deep-sea bed, creeping north in slow sheets. Antarctic Bottom Water flows many times as slowly as the water in surface currents: in the Pacific it may take a thousand years to reach the equator. Antarctic Bottom Water also flows into the Atlantic Ocean basin, where it flows north at a faster rate than in the Pacific. Antarctic Bottom Water has been identified as high as 40° N on the Atlantic floor.

A recording current meter

A small amount of dense bottom water also forms in the northern polar ocean. Although, the topography of the Arctic Ocean basin prevents most of the bottom water from escaping, with the exception of deep channels formed in the submarine ridges between Scotland, Iceland, and Greenland. These channels allow the cold, dense water formed in the Arctic to flow into the North Atlantic to form North Atlantic Deep Water. North Atlantic Deep Water forms when the relatively warm and salty North Atlantic Ocean cools as cold winds from northern Canada sweep over it. Exposed to the chilled air, water at the latitude of Iceland releases heat, cools from 10 °C to 2 °C, and sinks. Gulf Stream water that sinks in the north is replaced by warm water flowing clockwise along the U.S. east coast in the North Atlantic gyre.

Importance

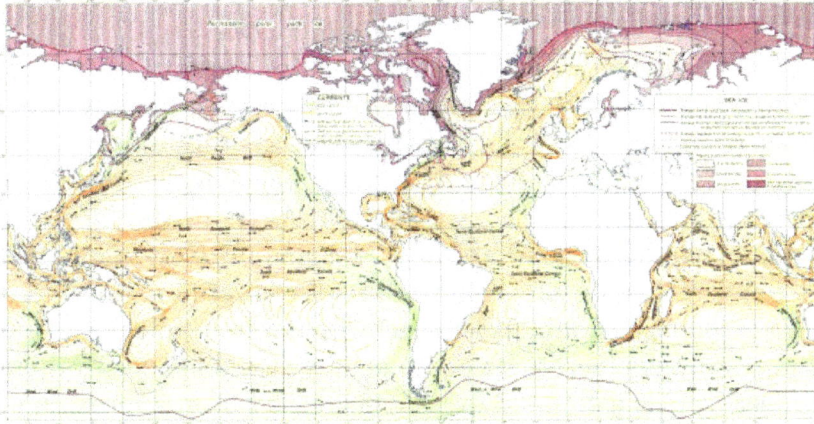

A 1943 map of the world's ocean currents.

Knowledge of surface ocean currents is essential in reducing costs of shipping, since traveling with them reduces fuel costs. In the wind powered sailing-ship era, knowledge was even more essential. A good example of this is the Agulhas Current, which long prevented Portuguese sailors from reaching India. In recent times, around-the-world sailing competitors make good use of surface currents to build and maintain speed. Ocean currents are also very important in the dispersal of many life forms. An example is the life-cycle of the European Eel.

Ocean currents are important in the study of marine debris, and vice versa. These currents also affect temperatures throughout the world. For example, the ocean current that brings warm water up the north Atlantic to northwest Europe also cumulatively and slowly blocks ice from forming along the seashores, which would also block ships from entering and exiting inland waterways and seaports, hence ocean currents play a decisive role in influencing the climates of regions through which they flow. Cold ocean water currents flowing from polar and sub-polar regions bring in a lot of plankton that are crucial to the continued survival of several key sea creature species in marine ecosystems. Since plankton are the food of fish, abundant fish populations often live where these currents prevail.

Ocean currents can also be used for marine power generation, with areas off of Japan, Florida and Hawaii being considered for test projects.

OSCAR: Near-realtime Global Ocean Surface Current Data Set

The OSCAR Near-realtime global ocean surface currents website from which users can create customized graphics and download the data. A section of the website provides validation studies in the form of graphics comparing OSCAR data with moored buoys and global drifters.

OSCAR data is used extensively in climate studies. maps and descriptions or annotations of climatic anomalies have been published in the monthly Climate Diagnostic Bulletin since 2001 and are routinely used to monitor ENSO and to test weather prediction models. OSCAR currents are routinely used to evaluate the surface currents in Global Circulation Models (GCMs), for example in NCEP Global Ocean Data Assimilation System (GODAS) and European Centre for Medium-Range Weather Forecasts (ECMWF).

Ocean Acidification

Ocean acidification is the ongoing decrease in the pH of the Earth's oceans, caused by the uptake of carbon dioxide (CO_2) from the atmosphere. Seawater is slightly basic (meaning pH > 7), and the process in question is a shift towards pH-neutral conditions rather than a transition to acidic conditions (pH < 7). Ocean alkalinity is not changed by the process, or may increase over long time periods due to carbonate dissolution. An estimated 30–40% of the carbon dioxide from human activity released into the atmosphere dissolves into oceans, rivers and lakes. To achieve chemical equilibrium, some of it reacts with the water to form carbonic acid. Some of these extra carbonic acid molecules react with a water molecule to give a bicarbonate ion and a hydronium ion, thus increasing ocean acidity (H^+ ion concentration). Between 1751 and 1994 surface ocean pH is estimated to have decreased from approximately 8.25 to 8.14, representing an increase of almost 30% in H^+ ion concentration in the world's oceans. Earth System Models project that within the last decade ocean acidity exceeded historical analogs and in combination with other ocean biogeochemical changes could undermine the functioning of marine ecosystems and disrupt the provision of many goods and services associated with the ocean.

Increasing acidity is thought to have a range of potentially harmful consequences for marine organisms, such as depressing metabolic rates and immune responses in some organisms, and causing coral bleaching. By increasing the presence of free hydrogen ions, each molecule of carbonic acid that forms in the oceans ultimately results in the conversion of *two* carbonate ions into bicarbonate ions. This net decrease in the amount of carbonate ions available makes it more difficult for marine calcifying organisms, such as coral and some plankton, to form biogenic calcium carbonate, and such structures become vulnerable to dissolution. Ongoing acidification of the oceans threatens food chains connected with the oceans. As members of the InterAcademy Panel, 105 science academies have issued a statement on ocean acidification recommending that by 2050, global CO_2 emissions be reduced by at least 50% compared to the 1990 level.

While ongoing ocean acidification is anthropogenic in origin, it has occurred previously in Earth's history. The most notable example is the Paleocene-Eocene Thermal Maximum (PETM), which occurred approximately 56 million years ago. For reasons that are currently uncertain, massive amounts of carbon entered the ocean and atmosphere, and led to the dissolution of carbonate sediments in all ocean basins.

Ocean acidification has been called the "evil twin of global warming" and "the other CO_2 problem".

Carbon Cycle

The carbon cycle describes the fluxes of carbon dioxide (CO 2) between the oceans, terrestrial biosphere, lithosphere, and the atmosphere. Human activities such as the combustion of fossil fuels and land use changes have led to a new flux of CO 2 into the atmosphere. About 45% has remained in the atmosphere; most of the rest has been taken up by the oceans, with some taken up by terrestrial plants.

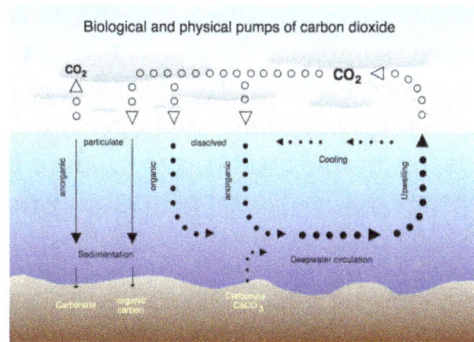

The CO 2 cycle between the atmosphere and the ocean

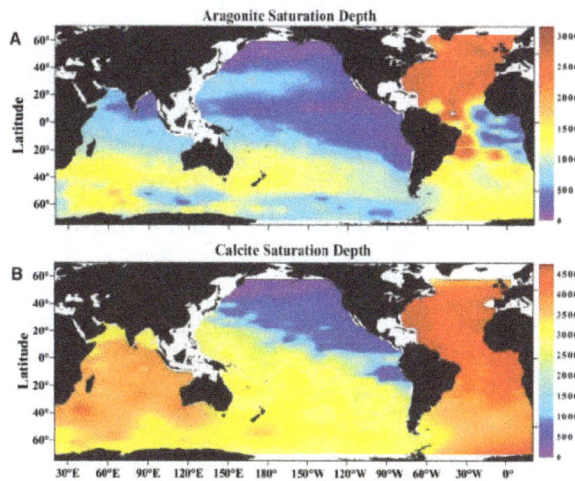

Distribution of (A) aragonite and (B) calcite saturation depth in the global oceans

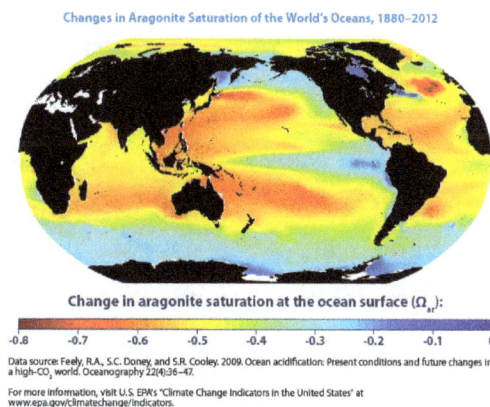

Changes in Aragonite Saturation of the World's Oceans, 1880–2012

Change in aragonite saturation at the ocean surface (Ω_{ar}):

| -0.8 | -0.7 | -0.6 | -0.5 | -0.4 | -0.3 | -0.2 | -0.1 | 0 |

Data source: Feely, R.A., S.C. Doney, and S.R. Cooley. 2009. Ocean acidification: Present conditions and future changes in a high-CO_2 world. Oceanography 22(4):36–47.

For more information, visit U.S. EPA's "Climate Change Indicators in the United States" at www.epa.gov/climatechange/indicators.

The map was created by the National Oceanic and Atmospheric Administration and the Woods Hole Oceanographic Institution using Community Earth System Model data. This map was created by comparing average conditions during the 1880s with average conditions during the most recent 10 years (2003–2012). Aragonite saturation has only been measured at selected locations during the last few decades, but it can be calculated reliably for different times and locations based on the relationships scientists have observed among aragonite saturation, pH, dissolved carbon, water temperature, concentrations of carbon dioxide in the atmosphere, and other factors that can be measured. This map shows changes in the amount of aragonite dissolved in ocean surface waters between the 1880s and the most recent decade (2003–2012). Aragonite saturation is a ratio that compares the amount of aragonite that is actually present with the total amount of aragonite that the water could hold if it were completely saturated. The more negative the change in aragonite saturation, the larger the decrease in aragonite available in the water, and the harder it is for marine creatures to produce their skeletons and shells. The global map shows changes over time in the amount of aragonite dissolved in ocean water, which is called aragonite saturation.

The carbon cycle involves both organic compounds such as cellulose and inorganic carbon compounds such as carbon dioxide and the carbonates. The inorganic compounds are particularly relevant when discussing ocean acidification for it includes many forms of dissolved CO_2 present in the Earth's oceans.

When CO_2 dissolves, it reacts with water to form a balance of ionic and non-ionic chemical species: dissolved free carbon dioxide ($CO_2(aq)$), carbonic acid (H_2CO_3), bicarbonate (HCO_3^-) and carbonate (CO_3^{2-}). The ratio of these species depends on factors such as seawater temperature and alkalinity (as shown in a Bjerrum plot). These different forms of dissolved inorganic carbon are transferred from an ocean's surface to its interior by the ocean's solubility pump.

The resistance of an area of ocean to absorbing atmospheric CO_2 is known as the Revelle factor.

Acidification

Dissolving CO_2 in seawater increases the hydrogen ion (H^+) concentration in the ocean, and thus decreases ocean pH, as follows:

$$CO_{2\ (aq)} + H_2O\ H_2CO_3\ HCO_3^- + H^+\ CO_3^{2-} + 2\ H^+.$$

Caldeira and Wickett (2003) placed the rate and magnitude of modern ocean acidification changes in the context of probable historical changes during the last 300 million years.

Since the industrial revolution began, it is estimated that surface ocean pH has dropped by slightly more than 0.1 units on the logarithmic scale of pH, representing about a 29% increase in H^+. It is expected to drop by a further 0.3 to 0.5 pH units (an additional doubling to tripling of today's post-industrial acid concentrations) by 2100 as the oceans absorb more anthropogenic CO_2, the impacts being most severe for coral reefs and the Southern Ocean. These changes are predicted to continue rapidly as the oceans take up more anthropogenic CO_2 from the atmosphere. The degree of change to ocean chemistry, including ocean pH, will depend on the mitigation and emissions pathways society takes.

Although the largest changes are expected in the future, a report from NOAA scientists found large quantities of water undersaturated in aragonite are already upwelling close to the Pacific continental shelf area of North America. Continental shelves play an important role in marine ecosystems since most marine organisms live or are spawned there, and though the study only dealt with the area from Vancouver to Northern California, the authors suggest that other shelf areas may be experiencing similar effects.

Average surface ocean pH				
Time	**pH**	**pH change relative to pre-industrial**	**Source**	**H⁺ concentration change relative to pre-industrial**
Pre-industrial (18th century)	8.179		analysed field	
Recent past (1990s)	8.104	−0.075	field	+ 18.9%
Present levels	~8.069	−0.11	field	+ 28.8%
2050 (2×CO2 = 560 ppm)	7.949	−0.230	model	+ 69.8%
2100 (IS92a)	7.824	−0.355	model	+ 126.5%

Rate

One of the first detailed datasets to examine how pH varied over a period of time at a temperate coastal location found that acidification was occurring much faster than previously predicted, with consequences for near-shore benthic ecosystems. Thomas Lovejoy, former chief biodiversity advisor to the World Bank, has suggested that "the acidity of the oceans will more than double in the next 40 years. This rate is 100 times faster than any changes in ocean acidity in the last 20 million years, making it unlikely that marine life can somehow adapt to the changes." It is predicted that, by the year 2100, the level of acidity in the ocean will reach the levels experienced by the earth 20 million years ago.

Current rates of ocean acidification have been compared with the greenhouse event at the Paleocene–Eocene boundary (about 55 million years ago) when surface ocean temperatures rose by 5–6 degrees Celsius. No catastrophe was seen in surface ecosystems, yet bottom-dwelling organisms in the deep ocean experienced a major extinction. The current acidification is on a path to reach levels higher than any seen in the last 65 million years, and the rate of increase is about ten times the rate that preceded the Paleocene–Eocene mass extinction. The current and projected acidification has been described as an almost unprecedented geological event. A National Research Council study released in April 2010 likewise concluded that "the level of acid in the oceans is increasing at an unprecedented rate." A 2012 paper in the journal *Science* examined the geological record in an attempt to find a historical analog for current global conditions as well as those of the future. The researchers determined that the current rate of ocean acidification is faster than at any time in the past 300 million years.

A review by climate scientists at the RealClimate blog, of a 2005 report by the Royal Society of the UK similarly highlighted the centrality of the *rates* of change in the present anthropogenic acidification process, writing:

"The natural pH of the ocean is determined by a need to balance the deposition and burial of CaCO3 on the sea floor against the influx of Ca2+and CO2−3 into the ocean from dissolving rocks on land, called weathering. These processes stabilize the pH of the ocean, by a mechanism called CaCO3 compensation...The point of bringing it up again is to note that if the CO2 concentration of the atmosphere changes more slowly than this, as it always has throughout the Vostok record, the pH of the ocean will be relatively unaffected because CaCO3 compensation can keep up. The [present] fossil fuel acidification is much faster than natural changes, and so the acid spike will be more intense than the earth has seen in at least 800,000 years."

In the 15-year period 1995–2010 alone, acidity has increased 6 percent in the upper 100 meters of the Pacific Ocean from Hawaii to Alaska. According to a statement in July 2012 by Jane Lubchenco, head of the U.S. National Oceanic and Atmospheric Administration "surface waters are changing much more rapidly than initial calculations have suggested. It's yet another reason to be very seriously concerned about the amount of carbon dioxide that is in the atmosphere now and the additional amount we continue to put out."

A 2013 study claimed acidity was increasing at a rate 10 times faster than in any of the evolutionary crises in Earth's history. In a synthesis report published in *Science* in 2015, 22 leading marine scientists stated that CO_2 from burning fossil fuels is changing the oceans' chemistry more rapidly than at any time since the Great Dying, Earth's most severe known extinction event, emphasizing that the 2 °C maximum temperature increase agreed upon by governments reflects too small a cut in emissions to prevent "dramatic impacts" on the world's oceans, with lead author Jean-Pierre Gattuso remarking that "The ocean has been minimally considered at previous climate negotiations. Our study provides compelling arguments for a radical change at the UN conference (in Paris) on climate change".

Calcification

Overview

Changes in ocean chemistry can have extensive direct and indirect effects on organisms and their habitats. One of the most important repercussions of increasing ocean acidity relates to the production of shells and plates out of calcium carbonate (CaCO3). This process is called calcification and is important to the biology and survival of a wide range of marine organisms. Calcification involves the precipitation of dissolved ions into solid CaCO3 structures, such as coccoliths. After they are formed, such structures are vulnerable to dissolution unless the surrounding seawater contains saturating concentrations of carbonate ions (CO_3^{2-}).

Mechanism

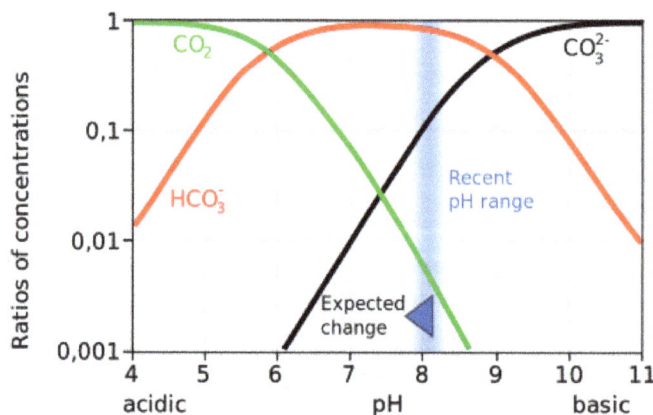

Bjerrum plot: Change in carbonate system of seawater from ocean acidification.

Of the extra carbon dioxide added into the oceans, some remains as dissolved carbon dioxide, while the rest contributes towards making additional bicarbonate (and additional carbonic acid). This also increases the concentration of hydrogen ions, and the percentage increase in hydrogen is larger than the percentage increase in bicarbonate, creating an imbalance in the reaction $HCO_3^- \rightleftharpoons CO_3^{2-} + H^+$. To maintain chemical equilibrium, some of the carbonate ions already in the ocean combine with some of the hydrogen ions to make further bicarbonate. Thus the ocean's concentration of carbonate ions is reduced, creating an imbalance in the reaction $Ca^{2+} + CO_3^{2-} \rightleftharpoons CaCO_3$, and making the dissolution of formed CaCO3 structures more likely.

These increases in concentrations of dissolved carbon dioxide and bicarbonate, and reduction in carbonate, are shown in a Bjerrum plot.

Saturation State

The saturation state (known as Ω) of seawater for a mineral is a measure of the thermodynamic potential for the mineral to form or to dissolve, and is described by the following equation:

$$\Omega = \frac{\left[Ca^{2+}\right]\left[CO_3^{2-}\right]}{K_{sp}}$$

Here Ω is the product of the concentrations (or activities) of the reacting ions that form the mineral (Ca2+ and CO2−3), divided by the product of the concentrations of those ions when the mineral is at equilibrium (Ksp), that is, when the mineral is neither forming nor dissolving. In seawater, a natural horizontal boundary is formed as a result of temperature, pressure, and depth, and is known as the saturation horizon, or lysocline. Above this saturation horizon, Ω has a value greater than 1, and CaCO3 does not readily dissolve. Most calcifying organisms live in such waters. Below this depth, Ω has a value less than 1, and CaCO3 will dissolve. However, if its production rate is high enough to offset dissolution, CaCO3 can still occur where Ω is less than 1. The carbonate compensation depth occurs at the depth in the ocean where production is exceeded by dissolution.

The decrease in the concentration of CO_3^{2-} decreases Ω, and hence makes CaCO3 dissolution more likely.

Calcium carbonate occurs in two common polymorphs (crystalline forms): aragonite and calcite. Aragonite is much more soluble than calcite, so the aragonite saturation horizon is always nearer to the surface than the calcite saturation horizon. This also means that those organisms that produce aragonite may be more vulnerable to changes in ocean acidity than those that produce calcite. Increasing CO2 levels and the resulting lower pH of seawater decreases the saturation state of CaCO3 and raises the saturation horizons of both forms closer to the surface. This decrease in saturation state is believed to be one of the main factors leading to decreased calcification in marine organisms, as the inorganic precipitation of CaCO3 is directly proportional to its saturation state.

Possible Impacts

Increasing acidity has possibly harmful consequences, such as depressing metabolic rates in jumbo squid, depressing the immune responses of blue mussels, and coral bleaching. However it may benefit some species, for example increasing the growth rate of the sea star, *Pisaster ochraceus*, while shelled plankton species may flourish in altered oceans.

Video summarizing the impacts of ocean acidification. Source: NOAA Environmental Visualization Laboratory.

The report "Ocean Acidification Summary for Policymakers 2013" describes research findings and possible impacts.

Impacts on Oceanic Calcifying Organisms

Although the natural absorption of CO2 by the world's oceans helps mitigate the climatic effects of anthropogenic emissions of CO2, it is believed that the resulting decrease in pH will have negative consequences, primarily for oceanic calcifying organisms. These span the food chain from autotrophs to heterotrophs and include organisms such as coccolithophores, corals, foraminifera, echinoderms, crustaceans and molluscs. As described above, under normal conditions, calcite and aragonite are stable in surface waters since the carbonate ion is at supersaturating concentrations. However, as ocean pH falls, the concentration of carbonate ions required for saturation to occur increases, and when carbonate becomes undersaturated, structures made of calcium carbonate are vulnerable to dissolution. Therefore, even if there is no change in the rate of calcification, the rate of dissolution of calcareous material increases.

Corals, coccolithophore algae, coralline algae, foraminifera, shellfish and pteropods experience reduced calcification or enhanced dissolution when exposed to elevated CO2.

The Royal Society published a comprehensive overview of ocean acidification, and its potential consequences, in June 2005. However, some studies have found different response to ocean acidification, with coccolithophore calcification and photosynthesis both increasing under elevated atmospheric pCO_2, an equal decline in primary production and calcification in response to elevated CO_2 or the direction of the response varying between species. A study in 2008 examining a sediment core from the North Atlantic found that while the species composition of coccolithophorids has remained unchanged for the industrial period 1780 to 2004, the calcification of coccoliths has increased by up to 40% during the same time. A 2010 study from Stony Brook University suggested that while some areas are overharvested and other fishing grounds are being restored, because of ocean acidification it may be impossible to bring back many previous shellfish populations. While the full ecological consequences of these changes in calcification are still uncertain, it appears likely that many calcifying species will be adversely affected.

When exposed in experiments to pH reduced by 0.2 to 0.4, larvae of a temperate brittlestar, a relative of the common sea star, fewer than 0.1 percent survived more than eight days. There is also a suggestion that a decline in the coccolithophores may have secondary effects on climate, contributing to global warming by decreasing the Earth's albedo via their effects on oceanic cloud cover. All marine ecosystems on Earth will be exposed to changes in acidification and several other ocean biogeochemical changes.

The fluid in the internal compartments where corals grow their exoskeleton is also extremely important for calcification growth. When the saturation rate of aragonite in the external seawater is at ambient levels, the corals will grow their aragonite crystals rapidly in their internal compartments, hence their exoskeleton grows rapidly. If the level of aragonite in the external seawater is lower than the ambient level, the corals have to work harder to maintain the right balance in the internal compartment. When that happens, the process of growing the crystals slows down, and this slows down the rate of how much their exoskeleton is growing. Depending on how much aragonite is in the surrounding water, the corals may even stop growing because the levels of aragonite are too low to pump in to the internal compartment. They could even dissolve faster than they can make the crystals to their skeleton, depending on the aragonite levels in the surrounding water.

Ocean acidification may force some organisms to reallocate resources away from productive endpoints such as growth in order to maintain calcification.

In some places carbon dioxide bubbles out from the sea floor, locally changing the pH and other aspects of the chemistry of the seawater. Studies of these carbon dioxide seeps have documented a variety of responses by different organisms. Coral reef communities located near carbon dioxide seeps are of particular interest because of the sensitivity of some corals species to acidification. In Papua New Guinea, declining pH caused by carbon dioxide seeps is associated with declines in coral species diversity. However, in Palau carbon dioxide seeps are not associated with reduced species diversity of corals, although bioerosion of coral skeletons is much higher at low pH sites.

Other Biological Impacts

Aside from the slowing and/or reversing of calcification, organisms may suffer other adverse effects, either indirectly through negative impacts on food resources, or directly as reproductive or physiological effects. For example, the elevated oceanic levels of CO_2 may produce CO 2-induced acidification of body fluids, known as hypercapnia. Also, increasing ocean acidity is believed to have a range of direct consequences. For example, increasing acidity has been observed to: reduce metabolic rates in jumbo squid; depress the immune responses of blue mussels; and make it harder for juvenile clownfish to tell apart the smells of non-predators and predators, or hear the sounds of their predators. This is possibly because ocean acidification may alter the acoustic properties of seawater, allowing sound to propagate further, and increasing ocean noise. Calcium carbonate ions are very important when it comes to building organisms as they use calcium to build skeletons and shells. OA affects ocean species to a varying degree. Many plants do well with high CO_2 levels, but organisms that calcify like clams, mussels, corals and sea urchins might not do so well because everything in the ocean is connected including the food web. Since OA is such a large event because the ocean takes up 71% of the earth's surface OA is being referred to as climate change because it will affect the whole planet if the ocean becomes acidic. This impacts all animals that use sound for echolocation or communication. Atlantic longfin squid eggs took longer to hatch in acidified water, and the squid's statolith was smaller and malformed in animals placed in sea water with a lower pH. The lower PH was simulated with 20-30 times the normal amount of CO_2.

However, as with calcification, as yet there is not a full understanding of these processes in marine organisms or ecosystems.

Another possible effect would be an increase in red tide events, which could contribute to the accumulation of toxins (domoic acid, brevetoxin, saxitoxin) in small organisms such as anchovies and shellfish, in turn increasing occurrences of amnesic shellfish poisoning, neurotoxic shellfish poisoning and paralytic shellfish poisoning.

Nonbiological Impacts

Leaving aside direct biological effects, it is expected that ocean acidification in the future will lead to a significant decrease in the burial of carbonate sediments for several centuries, and even the dissolution of existing carbonate sediments. This will cause an elevation of ocean alkalinity, leading to the enhancement of the ocean as a reservoir for CO_2 with implications for climate change as more CO_2 leaves the atmosphere for the ocean.

Impact on Human Industry

The threat of acidification includes a decline in commercial fisheries and in the Arctic tourism industry and economy. Commercial fisheries are threatened because acidification harms calcifying organisms which form the base of the Arctic food webs.

Pteropods and brittle stars both form the base of the Arctic food webs and are both seriously damaged from acidification. Pteropods shells dissolve with increasing acidification and the brittle stars lose muscle mass when re-growing appendages. For pteropods to create shells they require aragonite which is produced through carbonate ions and dissolved calcium. Pteropods are severely affected because increasing acidification levels have steadily decreased the amount of water supersaturated with carbonate which is needed for aragonite creation. Arctic waters are changing so rapidly that they will become undersaturated with aragonite as early as 2016. Additionally the brittle star's eggs die within a few days when exposed to expected conditions resulting from Arctic acidification. Acidification threatens to destroy Arctic food webs from the base up. Arctic food webs are considered simple, meaning there are few steps in the food chain from small organisms to larger predators. For example, pteropods are "a key prey item of a number of higher predators - larger plankton, fish, seabirds, whales" Both pteropods and sea stars serve as a substantial food source and their removal from the simple food web would pose a serious threat to the whole ecosystem. The effects on the calcifying organisms at the base of the food webs could potentially destroy fisheries. The value of fish caught from US commercial fisheries in 2007 was valued at $3.8 billion and of that 73% was derived from calcifiers and their direct predators. Other organisms are directly harmed as a result of acidification. For example, decrease in the growth of marine calcifiers such as the American Lobster, Ocean Quahog, and scallops means there is less shellfish meat available for sale and consumption. Red king crab fisheries are also at a serious threat because crabs are calcifiers and rely on carbonate ions for shell development. Baby red king crab when exposed to increased acidification levels experienced 100% mortality after 95 days. In 2006 Red King Cab accounted for 23% of the total guideline harvest levels and a serious decline in red crab population would threaten the crab harvesting industry. Several ocean goods and services are likely to be undermined by future ocean acidification potentially affecting the livelihoods of some 400 to 800 million people depending upon the emission scenario.

Impact on Indigenous Peoples

Acidification could damage the Arctic tourism economy and affect the way of life of indigenous peoples. A major pillar of Arctic tourism is the sport fishing and hunting industry. The sport fishing industry is threatened by collapsing food webs which provide food for the prized fish. A decline in tourism lowers revenue input in the area, and threatens the economies that are increasingly dependent on tourism. Acidification is not merely a threat but has significantly declined whole fish populations. For example, In Scandinavia studies conducted on acidic water revealed that 15% of species populations had disappeared and that many more populations were limited in numbers or declining. The rapid decrease or disappearance of marine life could also affect the diet of Indigenous peoples.

Possible Responses

Reducing CO_2 Emissions

Members of the InterAcademy Panel recommended that by 2050, global anthropogenic CO_2 emissions be reduced less than 50% of the 1990 level. The 2009 statement also called on world leaders to:

- Acknowledge that ocean acidification is a direct and real consequence of increasing atmospheric CO_2 concentrations, is already having an effect at current concentrations, and is likely to cause grave harm to important marine ecosystems as CO_2 concentrations reach 450 [parts-per-million (ppm)] and above;

- [...] Recognise that reducing the build up of CO_2 in the atmosphere is the only practicable solution to mitigating ocean acidification;

- [...] Reinvigorate action to reduce stressors, such as overfishing and pollution, on marine ecosystems to increase resilience to ocean acidification.

Stabilizing atmospheric CO_2 concentrations at 450 ppm would require near-term emissions reductions, with steeper reductions over time.

The German Advisory Council on Global Change stated:

In order to prevent disruption of the calcification of marine organisms and the resultant risk of fundamentally altering marine food webs, the following guard rail should be obeyed: the pH of near surface waters should not drop more than 0.2 units below the pre-industrial average value in any larger ocean region (nor in the global mean).

One policy target related to ocean acidity is the magnitude of future global warming. Parties to the United Nations Framework Convention on Climate Change (UNFCCC) adopted a target of limiting warming to below 2 °C, relative to the pre-industrial level. Meeting this target would require substantial reductions in anthropogenic CO_2 emissions.

Limiting global warming to below 2 °C would imply a reduction in surface ocean pH of 0.16 from pre-industrial levels. This would represent a substantial decline in surface ocean pH.

Climate Engineering

Climate engineering (mitigating temperature or pH effects of emissions) has been proposed as a possible response to ocean acidification. The IAP (2009) statement cautioned against climate engineering as a policy response:

Mitigation approaches such as adding chemicals to counter the effects of acidification are likely to be expensive, only partly effective and only at a very local scale, and may pose additional unanticipated risks to the marine environment. There has been very little research on the feasibility and impacts of these approaches. Substantial research is needed before these techniques could be applied.

Reports by the WGBU (2006), the UK's Royal Society (2009), and the US National Research Council (2011) warned of the potential risks and difficulties associated with climate engineering.

Iron Fertilization

Iron fertilization of the ocean could stimulate photosynthesis in phytoplankton. The phytoplankton would convert the ocean's dissolved carbon dioxide into carbohydrate and oxygen gas, some of which would sink into the deeper ocean before oxidizing. More than a dozen open-sea experiments confirmed that adding iron to the ocean increases photosynthesis in phytoplankton by up to 30 times. While this approach has been proposed as a potential solution to the ocean acidification problem, mitigation of surface ocean acidification might increase acidification in the less-inhabited deep ocean.

A report by the UK's Royal Society (2009) reviewed the approach for effectiveness, affordability, timeliness and safety. The rating for affordability was "medium", or "not expected to be very cost-effective." For the other three criteria, the ratings ranged from "low" to "very low" (i.e., not good). For example, in regards to safety, the report found a "[high] potential for undesirable ecological side effects," and that ocean fertilization "may increase anoxic regions of ocean ('dead zones')."

Carbon Negative Fuels

Carbonic acid can be extracted from seawater as carbon dioxide for use in making synthetic fuel. If the resulting flue exhaust gas was subject to carbon capture, then the process would be carbon negative over time, resulting in permanent extraction of inorganic carbon from seawater and the atmosphere with which seawater is in equilibrium. Based on the energy requirements, this process was estimated to cost about $50 per tonne of CO_2.

Ocean Heat Content

Oceanic heat content (OHC) is the heat stored in the ocean. Oceanography and climatology are the science branches which study ocean heat content. Changes in the ocean heat content play an important role in the sea level rise, because of thermal expansion. It is with high confidence that ocean warming accounts for 90% of the energy accumulation from global warming between 1971 and 2010.

Definition and Measurement

The areal density of ocean heat content between two depth levels is defined as:

$$H = \rho c_p \int_{h2}^{h1} T(z)dz$$

Where ρ is seawater density, c_p is the specific heat of sea water, h2 is the lower depth, h1 is the upper depth, and $T(z)$ is the temperature profile. In SI units, H has units of $J \cdot m^{-2}$. Multiplying this quantity by the area of an ocean basin, or entire ocean, gives the total heat content, as indicated in the figure to right.

Ocean heat content can be computed using temperature measurements obtained by a Nansen bottle, an ARGO float, or ocean acoustic tomography. The World Ocean Database Project is the largest database for temperature profiles from all of the world's oceans.

Recent Changes

Several studies in recent years have found a multidecadal increase in OHC of the deep and upper ocean regions and attribute the heat uptake to anthropogenic warming. Studies based on *ARGO* indicate that ocean surface winds, especially the subtropical trade winds in the Pacific Ocean, change ocean heat vertical distribution. This results in changes among ocean currents, and an increase of the subtropical overturning, which is also related to the El Niño and La Niña phenomenon. Depending on stochastic natural variability fluctuations, during La Niña years around 30% more heat from the upper ocean layer is transported into the deeper ocean. Model studies indicate that ocean currents transport more heat into deeper layers during La Niña years, following changes in wind circulation. Years with increased ocean heat uptake have been associated with negative phases of the interdecadal Pacific oscillation (IPO). This is of particular interest to climate scientists who use the data to estimate the *ocean heat uptake*.

A study in 2015 concluded that ocean heat content increases by the Pacific Ocean, were compensated by an abrupt distribution of OHC into the Indian Ocean.

Role of Ocean Heat Content in Sea level Rise

Sea Level Rise

Sea level rise has been estimated to be on average between +2.6 millimetres (0.10 in) and 2.9 millimetres (0.11 in) per year ± 0.4 millimetres (0.016 in) since 1993.

Additionally, sea level rise has accelerated in recent years. For the period between 1870 and 2004, global average sea levels are estimated to have risen a total of 195 millimetres (7.7 in), and 1.7 millimetres (0.067 in) ± 0.3 millimetres (0.012 in) per year, with a significant acceleration of sea-level rise of 0.013 millimetres (0.00051 in) ± 0.006 millimetres (0.00024 in) per year per year.

According to one study of measurements available from 1950 to 2009, these measurements show an average annual rise in sea level of 1.7 millimetres (0.067 in) ± 0.3 millimetres (0.012 in) per year during this period, with satellite data showing a rise of 3.3 millimetres (0.13 in) ± 0.4 millimetres (0.016 in) per year from 1993 to 2009. Sea level rise is one of several lines of evidence that support the view that the global climate has recently warmed. In 2014 the USGCRP National

Climate Assessment projected that by the year 2100, the average sea level rise will have been between one and four feet (300mm-1200mm) since the date of the 2014 assessment. Current rates of sea level rise have roughly doubled since the pre 1992 rates of sea level rise of the 20th century.

In 2007, the Intergovernmental Panel on Climate Change (IPCC) stated that it is very likely human-induced (anthropogenic) warming contributed to the sea level rise observed in the latter half of the 20th century. The 2013 IPCC report (AR5) concluded, *"there is high confidence that the rate of sea level rise has increased during the last two centuries, and it is likely that GMSL (Global Mean Sea Level) has accelerated since the early 1900's.*

Sea level rises can considerably influence human populations in coastal and island regions and natural environments like marine ecosystems. Sea level rise is expected to continue for centuries. Because of the slow inertia, long response time for parts of the climate system, it has been estimated that we are already committed to a sea-level rise of approximately 2.3 metres (7.5 ft) for each degree Celsius of temperature rise within the next 2,000 years. It has been suggested that besides CO2 emissions reductions, a short term action to reduce sea level rise is to cut emissions of heat trapping gases such as methane and particulates such as soot.

Mechanism

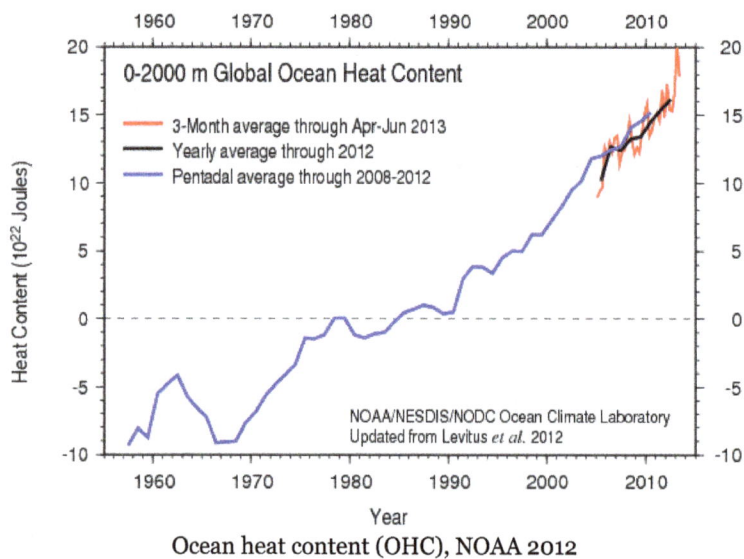

Ocean heat content (OHC), NOAA 2012

There are two main mechanisms that contribute to observed sea level rise: (1) thermal expansion: because of the increase in ocean heat content (ocean water expands as it warms); and (2) the melting of major stores of land ice like ice sheets and glaciers.

On the timescale of centuries to millennia, the melting of ice sheets could result in even higher sea level rise. Partial deglaciation of the Greenland ice sheet, and possibly the West Antarctic ice sheet, could contribute 4 to 6 m (13 to 20 ft) or more to sea level rise.

Past Changes in Sea Level

Various factors affect the volume or mass of the ocean, leading to long-term changes in eustatic sea level. The two primary influences are temperature (because the density of water depends on temperature), and the mass of water locked up on land and sea as fresh water in rivers, lakes, glaciers and polar ice caps. Over much longer geological timescales, changes in the shape of oceanic basins and in land–sea distribution affect sea level. Since the Last Glacial Maximum about 20,000 years ago, sea level has risen by more than 125 m, with rates varying from tenths of a mm/yr to 10+mm/year, as a result of melting of major ice sheets.

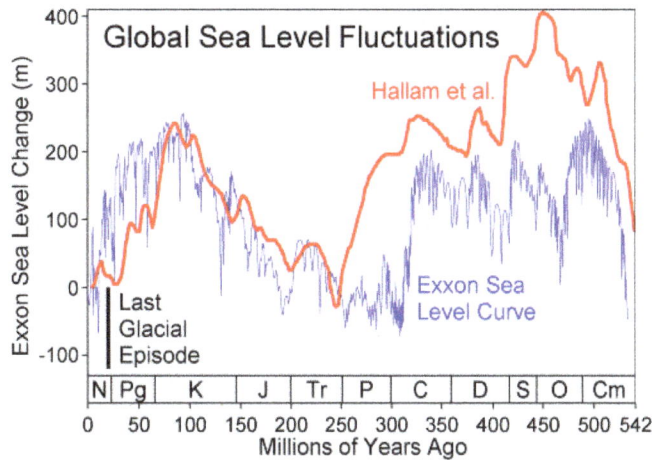

Comparison of two sea level reconstructions during the last 500 Ma. The scale of change during the last glacial/interglacial transition is indicated with a black bar. Note that over most of geologic history, long-term average sea level has been significantly higher than today.

During deglaciation between about 19,000 and 8,000 calendar years ago, sea level rose at extremely high rates as the result of the rapid melting of the British-Irish Sea, Fennoscandian, Laurentide, Barents-Kara, Patagonian, Innuitian ice sheets and parts of the Antarctic ice sheet. At the onset of deglaciation about 19,000 calendar years ago, a brief, at most 500-year long, glacio-eustatic event may have contributed as much as 10 m to sea level with an average rate of about 20 mm/yr. During the rest of the early Holocene, the rate of sea level rise varied from a low of about 6.0 - 9.9 mm/yr to as high as 30 - 60 mm/yr during brief periods of accelerated sea level rise.

Solid geological evidence, based largely upon analysis of deep cores of coral reefs, exists only for 3 major periods of accelerated sea level rise, called *meltwater pulses*, during the last deglaciation. They are Meltwater pulse 1A between circa 14,600 and 14,300 calendar years ago; Meltwater pulse 1B between circa 11,400 and 11,100 calendar years ago; and Meltwater pulse 1C between 8,200 and 7,600 calendar years ago. Meltwater pulse 1A was a 13.5 m rise over about 290 years centered at 14,200 calendar years ago and Meltwater pulse 1B was a 7.5 m rise over about 160 years centered at 11,000 years calendar years ago. In sharp contrast, the period between 14,300 and 11,100 calendar years ago, which includes the Younger Dryas interval, was an interval of reduced sea level rise at about 6.0 - 9.9 mm/yr. Meltwater pulse 1C was centered at 8,000 calendar years and produced a rise of 6.5 m in less than 140 year. Such rapid rates of sea level rising during meltwater events clearly implicate major ice-loss events related to ice sheet collapse. The primary source may have been meltwater from the Antarctic ice sheet. Other studies suggest a Northern Hemisphere source for the meltwater in the Laurentide ice sheet.

Recently, it has become widely accepted that late Holocene, 3,000 calendar years ago to present, sea level was nearly stable prior to an acceleration of rate of rise that is variously dated between

1850 and 1900 AD. Late Holocene rates of sea level rise have been estimated using evidence from archaeological sites and late Holocene tidal marsh sediments, combined with tide gauge and satellite records and geophysical modeling. For example, this research included studies of Roman wells in Caesarea and of Roman *piscinae* in Italy. These methods in combination suggest a mean eustatic component of 0.07 mm/yr for the last 2000 years.

Since 1880, as the Industrial Revolution took center stage, the ocean began to rise briskly, climbing a total of 210 mm (8.3 in) through 2009 causing extensive erosion worldwide and costing billions.

Sea level rose by 6 cm during the 19th century and 19 cm in the 20th century. Evidence for this includes geological observations, the longest instrumental records and the observed rate of 20th century sea level rise. For example, geological observations indicate that during the last 2,000 years, sea level change was small, with an average rate of only 0.0–0.2 mm per year. This compares to an average rate of 1.7 ± 0.5 mm per year for the 20th century. Baart et al. (2012) show that it is important to account for the effect of the 18.6-year lunar nodal cycle before acceleration in sea level rise should be concluded. Based on tide gauge data, the rate of global average sea level rise during the 20th century lies in the range 0.8 to 3.3 mm/yr, with an average rate of 1.8 mm/yr.

A two degrees Celsius of warming would warm the Earth above Eemian levels, move conditions closer to the Pliocene climate, a time when sea level was in the range of 25 meters higher than today. However, one study argues that sea level during the Pliocene might have only risen by 9 to 13.5 meters, due to more resilient ice sheets. Warren Cornwall, in: 'Ghosts of Ocean Past', published in an 'Science' monographic issue, 13 November 2015: 'Sea changes', pgs 752-755, presented a chart showing the current warming respect to preindustrial era of 1 °C, that goes along with the current CO_2 in atmosphere of 400 ppm. With the same 400 ppm CO_2, 3 million years ago, with an increased average temperature of 2 to 3 °C above our preindustrial levels, Sea level was between 6 meters and a not defined enough upper range. The issue may be not if Sea Level will raise, but how much, and at what a pace.

Projections

This graph shows the projected change in global sea level rise if atmospheric carbon dioxide (CO_2) concentrations were to either quadruple or double. The projection is based on several multi-century integrations of a GFDL global coupled ocean-atmosphere model. These projections are the expected changes due to thermal expansion of sea water alone, and do not include the effect of melted continental ice sheets. With the effect of ice sheets included, the total rise could be larger by a substantial factor. Image credit: NOAA GFDL.

21st Century

The 2007 Fourth Assessment Report (IPCC 4) projected century-end sea levels using the Special Report on Emissions Scenarios (SRES). SRES developed emissions scenarios to project climate-change impacts. The projections based on these scenarios are not predictions, but reflect plausible estimates of future social and economic development (e.g., economic growth, population level). The six SRES "marker" scenarios projected sea level to rise by 18 to 59 centimetres (7.1 to 23.2 in).

Their projections were for the time period 2090–99, with the increase in level relative to average sea level over the 1980–99 period. This estimate did not include all of the possible contributions of ice sheets.

Hansen (2007), assumed an ice sheet contribution of 1 cm for the decade 2005–15, with a potential ten year doubling time for sea-level rise, based on a nonlinear ice sheet response, which would yield 5 m this century.

Research from 2008 observed rapid declines in ice-mass balance from both Greenland and Antarctica, and concluded that sea-level rise by 2100 is likely to be at least twice as large as that presented by IPCC AR4, with an upper limit of about two meters.

Projections assessed by the US National Research Council (2010) suggest possible sea level rise over the 21st century of between 56 and 200 cm (22 and 79 in). The NRC describes the IPCC projections as "conservative".

In 2011, Rignot and others projected a rise of 32 centimetres (13 in) by 2050. Their projection included increased contributions from the Antarctic and Greenland ice sheets. Use of two completely different approaches reinforced the Rignot projection.

In its Fifth Assessment Report (2013), The IPCC found that recent observations of global average sea level rise at a rate of 3.2 [2.8 to 3.6] mm per year is consistent with the sum of contributions from observed thermal ocean expansion due to rising temperatures (1.1 [0.8 to 1.4] mm per year), glacier melt (0.76 [0.39 to 1.13] mm per year), Greenland ice sheet melt (0.33 [0.25 to 0.41] mm per year), Antarctic ice sheet melt (0.27 [0.16 to 0.38] mm per year), and changes to land water storage (0.38 [0.26 to 0.49] mm per year). The report had also concluded that if emissions continue to keep up with the worst case IPCC scenarios, global average sea level could rise by nearly 1m by 2100 (0.52–0.98 m from a 1986-2005 baseline). If emissions follow the lowest emissions scenario, then global average sea level is projected to rise by between 0.28–0.6 m by 2100 (compared to a 1986–2005 baseline).

The Third National Climate Assessment (NCA), released May 6, 2014, projected a sea level rise of 1 to 4 feet by 2100 (30–120 cm). Decision makers who are particularly susceptible to risk may wish to use a wider range of scenarios from 8 inches to 6.6 feet by 2100.

A 2015 study by sea level rise experts concluded that based on MIS 5e data, sea level rise could rise faster in the coming decades, with a doubling time of 10, 20 or 40 years. The study abstract explains: *We argue that ice sheets in contact with the ocean are vulnerable to non-linear disintegration in response to ocean warming, and we posit that ice sheet mass loss can be approximated by a doubling time up to sea level rise of at least several meters. Doubling times of 10, 20 or 40 years yield sea level rise of several meters in 50, 100 or 200 years. Paleoclimate data reveal that subsurface ocean warming causes ice shelf melt and ice sheet discharge.*

Our climate model exposes amplifying feedbacks in the Southern Ocean that slow Antarctic bottom water formation and increase ocean temperature near ice shelf grounding lines, while cooling the surface ocean and increasing sea ice cover and water column stability. Ocean surface cooling, in the North Atlantic as well as the Southern Ocean, increases tropospheric horizontal temperature gradients, eddy kinetic energy and baroclinicity, which drive more powerful storms. However, Greg Holland from the National Center for Atmospheric Research, who reviewed the study, noted *"There is no doubt that the sea level rise, within the IPCC, is a very conservative number, so the truth lies somewhere between IPCC and Jim."*

After 2100

There is a widespread consensus that substantial long-term sea-level rise will continue for centuries to come. IPCC AR4 estimated that at least a partial deglaciation of the Greenland ice sheet, and possibly the West Antarctic ice sheet, would occur given a global average temperature increase of 1–4 °C (relative to temperatures over the years 1990–2000). This estimate was given about a 50% chance of being correct. The estimated timescale was centuries to millennia, and would contribute 4 to 6 metres (13 to 20 ft) or more to sea levels over this period.

Models

There is the possibility of a rapid change in glaciers, ice sheets, and hence sea level. Predictions of such a change are highly uncertain due to a lack of scientific understanding. Modeling of the processes associated with a rapid ice-sheet and glacier change could potentially increase future projections of sea-level rise.

Hansen (2007), concluded that paleoclimate ice sheet models generally do not include physics of ice streams, effects of surface melt descending through crevasses and lubricating basal flow, or realistic interactions with the ocean. The calibration of projected modelling for future sea-level rise is generally done with a linear projection of future sea level. Thus, does not include potential nonlinear collapse of an ice sheet.

Contribution

Close-up of Ross Ice Shelf, the largest ice shelf of Antarctica, about the size of France and up to several hundred metres thick.

Each year about 8 mm of precipitation (liquid equivalent) falls on the ice sheets in Antarctica and Greenland, mostly as snow, which accumulates and over time forms glacial ice. Much of this precipitation began as water vapor evaporated from the ocean surface. To a first approximation, the same amount of water appeared to return to the ocean in icebergs and from ice melting at the edges. Scientists previously had estimated which is greater, ice going in or coming out, called the mass balance, important because a nonzero balance causes changes in global sea level. High-precision gravimetry from satellites determined that Greenland was losing more than 200 billion tons of ice per year, in accord with loss estimates from ground measurement. The rate of ice loss was accelerating, having grown from 137 gigatons in 2002–2003.

- The total global ice mass lost from Greenland, Antarctica and Earth's glaciers and ice caps during 2003–2010 was about 4.3 trillion tons (1,000 cubic miles), adding about 12 mm (0.5 in) to global sea level, enough ice to cover an area comparable to the United States 50 cm (1.5 ft) deep.

- The melting of small glaciers on the margins of Greenland and the Antarctic Peninsula would increase sea level around 0.5 meter. At the extreme potential, according to the Third Assessment Report of the International Panel on Climate Change, the ice contained within the Greenland ice sheet entirely melted increases sea level by 7.2 meters (24 feet). The ice contained within the Antarctic ice sheet entirely melted would produce 61.1 meters (200 feet) of sea-level change, both totaling a sea-level rise of 68.3 meters (224 feet).

It is estimated that Antarctica, if fully melted, would contribute more than 60 metres of sea level rise, and Greenland would contribute more than 7 metres. Small glaciers and ice caps on the margins of Greenland and the Antarctic Peninsula might contribute about 0.5 metres. While the latter figure is much smaller than for Antarctica or Greenland it could occur relatively quickly (within the coming century) whereas melting of Greenland would be slow (perhaps 1,500 years to fully deglaciate at the fastest likely rate) and Antarctica even slower. However, this calculation does not account for the possibility that as meltwater flows under and lubricates the larger ice sheets, they could begin to move much more rapidly towards the sea.

In 2002, Rignot and Thomas found that the West Antarctic and Greenland ice sheets were losing mass, while the East Antarctic ice sheet was probably in balance (although they could not determine the sign of the mass balance for The East Antarctic ice sheet). Kwok and Comiso (*J. Climate*, v15, 487–501, 2002) also discovered that temperature and pressure anomalies around West Antarctica and on the other side of the Antarctic Peninsula correlate with recent Southern Oscillation events.

In 2005 it was reported that during 1992–2003, East Antarctica thickened at an average rate of about 18 mm/yr while West Antarctica showed an overall thinning of 9 mm/yr. associated with increased precipitation. A gain of this magnitude is enough to slow sea-level rise by 0.12 ± 0.02 mm/yr.

Antarctica

On the Antarctic continent itself, the large volume of ice present stores around 70% of the world's fresh water. This ice sheet is constantly gaining ice from snowfall and losing ice through outflow to the sea.

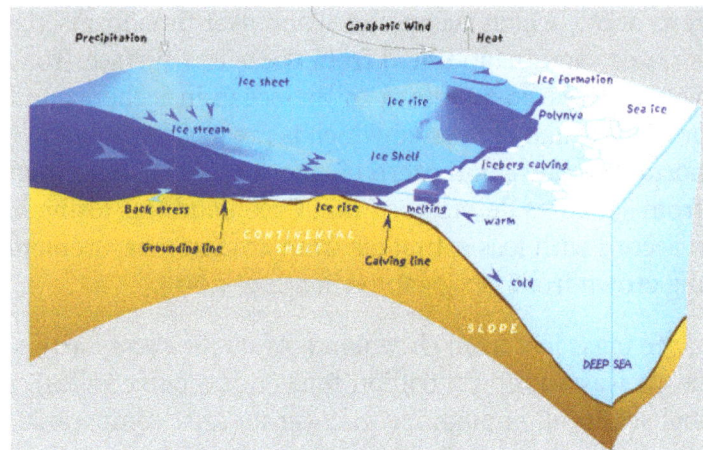

Processes around an Antarctic ice shelf

Sheperd et al. 2012, found that different satellite methods were in good agreement and combing methods leads to more certainty with East Antarctica, West Antarctica, and the Antarctic Peninsula changing in mass by +14 ± 43, −65 ± 26, and −20 ± 14 gigatonnes per year.

East Antarctic Ice Sheet (EAIS)

East Antarctica is a cold region with a ground-base above sea level and occupies most of the continent. This area is dominated by small accumulations of snowfall which becomes ice and thus eventually seaward glacial flows. The mass balance of the East Antarctic Ice Sheet as a whole over the period 1980-2004 is thought to be slightly positive (lowering sea level) or near to balance, with a large degree of uncertainty. However, increased ice outflow has been suggested in some regions.

West Antarctic Ice Sheet (WAIS)

West Antarctica is currently experiencing a net outflow of glacial ice, which will increase global sea level over time. A review of the scientific studies looking at data from 1992 to 2006 suggested a net loss of around 50 gigatons of ice per year was a reasonable estimate (around 0.14 mm of sea-level rise), although significant acceleration of outflow glaciers in the Amundsen Sea Embayment could have more than doubled this figure for the year 2006.

Thomas et al. found evidence of an accelerated contribution to sea level rise from West Antarctica. The data showed that the Amundsen Sea sector of the West Antarctic Ice Sheet was discharging 250 cubic kilometres of ice every year, which was 60% more than precipitation accumulation in the catchment areas. This alone was sufficient to raise sea level at 0.24 mm/yr. Further, thinning rates for the glaciers studied in 2002–03 had increased over the values measured in the early 1990s. The bedrock underlying the glaciers was found to be hundreds of metres deeper than previously known, indicating exit routes for ice from further inland in the Byrd Subpolar Basin. Thus the West Antarctic ice sheet may not be as stable as has been supposed.

A 2009 study found that the rapid collapse of West Antarctic Ice Sheet would raise sea level by 3.3 metres (11 ft).

Glaciers

Observational and modelling studies of mass loss from glaciers and ice caps indicate a contribution to sea-level rise of 0.2–0.4 mm/yr, averaged over the 20th century. The results from Dyurgerov show a sharp increase in the contribution of mountain and subpolar glaciers to sea-level rise since 1996 (0.5 mm/yr) to 1998 (2 mm/yr) with an average of about 0.35 mm/yr since 1960. Of interest also is Arendt et al., who estimate the contribution of Alaskan glaciers of 0.14±0.04 mm/yr between the mid-1950s to the mid-1990s, increasing to 0.27 mm/yr in the middle and late 1990s.

Greenland

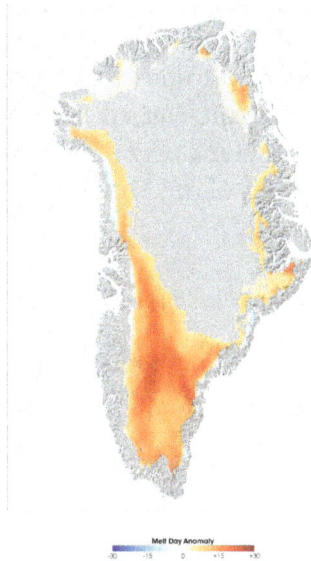

Greenland 2007 melt anomaly, measured as the difference between the number of days on which melting occurred in 2007 compared to the average annual melting days from 1988–2006

In 2004 Rignot et al. estimated a contribution of 0.04 ± 0.01 mm/yr to sea level rise from South East Greenland. In the same year, Krabill *et al.* estimate a net contribution from Greenland to be at least 0.13 mm/yr in the 1990s. Joughin *et al.* have measured a doubling of the speed of Jakobshavn Isbræ between 1997 and 2003. This is Greenland's largest outlet glacier; it drains 6.5% of the ice sheet, and is thought to be responsible for increasing the rate of sea-level rise by about 0.06 millimetres per year, or roughly 4% of the 20th-century rate of sea-level increase. In 2004, Rignot *et al.* estimated a contribution of 0.04±0.01 mm/yr to sea-level rise from southeast Greenland.

Rignot and Kanagaratnam produced a comprehensive study and map of the outlet glaciers and basins of Greenland. They found widespread glacial acceleration below 66 N in 1996 which spread to 70 N by 2005; and that the ice sheet loss rate in that decade increased from 90 to 200 cubic km/ yr; this corresponds to an extra 0.25–0.55 mm/yr of sea level rise.

In July 2005 it was reported that the Kangerlussuaq Glacier, on Greenland's east coast, was moving towards the sea three times faster than a decade earlier. Kangerdlugssuaq is around 1,000 m thick, 7.2 km (4.5 miles) wide, and drains about 4% of the ice from the Greenland ice sheet. Measurements of Kangerdlugssuaq in 1988 and 1996 showed it moving at between 5 and 6 km/yr (3.1–3.7 miles/yr), while in 2005 that speed had increased to 14 km/yr (8.7 miles/yr).

According to the 2004 Arctic Climate Impact Assessment, climate models project that local warming in Greenland will exceed 3 °C during this century. Also, ice-sheet models project that such a warming would initiate the long-term melting of the ice sheet, leading to a complete melting of the Greenland ice sheet over several millennia, resulting in a global sea level rise of about seven metres.

Subsidence and Effective Sea Level Rise

Many ports, urban conglomerations, and agricultural regions are built on river deltas, where subsidence contributes to a substantial increase in *effective* sea level rise. This is caused by both unsustainable extraction of groundwater (in some place also by extraction of oil and gas), and by levees and other flood management practices that prevent accumulation of sediments to compensate for the natural settling of deltaic soils. In many deltas this results in subsidence ranging from several millimeters per year up to possibly 25 centimeters per year in parts of the Ciliwung delta (Jakarta). Total anthropogenic-caused subsidence in the Rhine-Meuse-Scheldt delta (Netherlands) is estimated at 3 to 4 meters, over nine meters in the Sacramento-San Joaquin River Delta, and over ten feet in urban areas of the Mississippi River Delta (New Orleans).

Effects

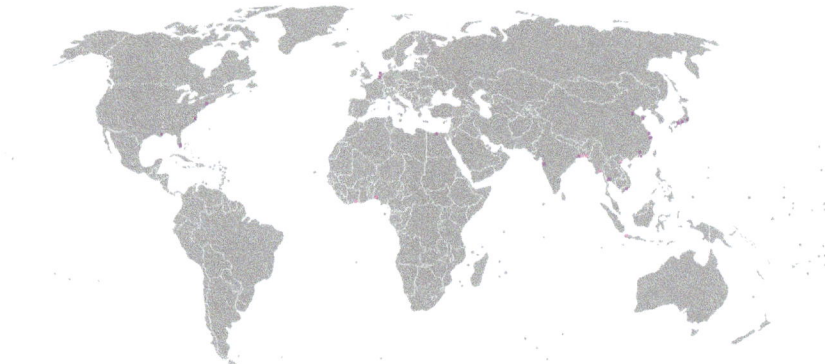

Map of major cities of the world most vulnerable to sea level rise

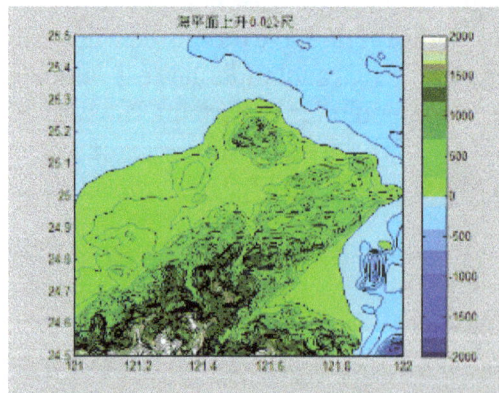

Schematic animation of sea level rise in Taipei, Taiwan and surrounding regions, in meters

The IPCC TAR WGII report (*Impacts, Adaptation Vulnerability*) notes that current and future climate change would be expected to have a number of impacts, particularly on coastal systems. Such impacts may include increased coastal erosion, higher storm-surge flooding, inhibition of primary production processes, more extensive coastal inundation, changes in surface water quality and groundwater characteristics, increased loss of property and coastal habitats, increased flood risk and potential loss of life, loss of non-monetary cultural resources and values, impacts on agriculture and aquaculture through decline in soil and water quality, and loss of tourism, recreation, and transportation functions.

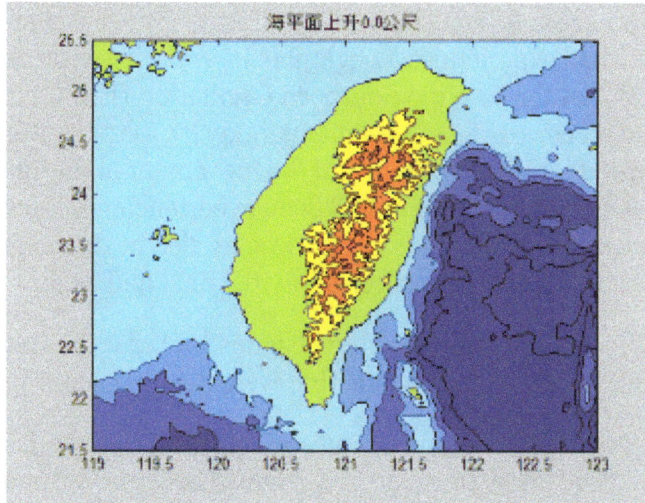

Schematic animation of sea level rise in Taiwan and surrounding regions, in meters

There is an implication that many of these impacts will be detrimental—especially for the three-quarters of the world's poor who depend on agriculture systems. The report does, however, note that owing to the great diversity of coastal environments; regional and local differences in projected relative sea level and climate changes; and differences in the resilience and adaptive capacity of ecosystems, sectors, and countries, the impacts will be highly variable in time and space.

The IPCC report of 2007 estimated that accelerated melting of the Himalayan ice caps and the resulting rise in sea levels would likely increase the severity of flooding in the short term during the rainy season and greatly magnify the impact of tidal storm surges during the cyclone season. A sea-level rise of just 400 mm in the Bay of Bengal would put 11 percent of the Bangladesh's coastal land underwater, creating 7–10 million climate refugees.

Sea level rise could also displace many shore-based populations: for example it is estimated that a sea level rise of just 200 mm could make 740,000 people in Nigeria homeless.

Future sea-level rise, like the recent rise, is not expected to be globally uniform. Some regions show a sea-level rise substantially more than the global average (in many cases of more than twice the average), and others a sea level fall. However, models disagree as to the likely pattern of sea level change.

Island Nations

IPCC assessments suggest that deltas and small island states are particularly vulnerable to sea-level rise caused by both thermal expansion and increased ocean water. Sea level changes have not yet been conclusively proven to have directly resulted in environmental, humanitarian, or economic losses to small island states, but the IPCC and other bodies have found this a serious risk scenario in coming decades.

Maldives, Tuvalu, and other low-lying countries are among the areas that are at the highest level of risk. The UN's environmental panel has warned that, at current rates, sea level would be high enough to make the Maldives uninhabitable by 2100.

Many media reports have focused on the island nations of the Pacific, notably the Polynesian islands of Tuvalu, which based on more severe flooding events in recent years, were thought to be "sinking" due to sea level rise. A scientific review in 2000 reported that based on University of Hawaii gauge data, Tuvalu had experienced a negligible increase in sea level of 0.07 mm a year over the past two decades, and that the El Niño Southern Oscillation (ENSO) had been a larger factor in Tuvalu's higher tides in recent years. A subsequent study by John Hunter from the University of Tasmania, however, adjusted for ENSO effects and the movement of the gauge (which was thought to be sinking). Hunter concluded that Tuvalu had been experiencing sea-level rise of about 1.2 mm per year. The recent more frequent flooding in Tuvalu may also be due to an erosional loss of land during and following the actions of 1997 cyclones Gavin, Hina, and Keli.

In 2016 it was reported that five of the Solomon Islands had disappeared due to the combined effects of sea level rise and stronger trade winds that were pushing water into the Western Pacific.

Besides the issues that flooding brings, such as soil salinisation, the island states themselves would also become dissolved over time, as the islands become uninhabitable or completely submerged by the sea. Once this happens, all rights on the surrounding area (sea) are removed. This area can be huge as rights extend to a radius of 224 nautical miles (414 km) around the entire island state. Any resources, such as fossil oil, minerals and metals, within this area can be freely dug up by anyone and sold without needing to pay any commission to the (now dissolved) island state.

Options that have been proposed to assist island nations to adapt to rising sea level include abandoning islands, building dikes, and building upwards.

Cities

A study in the April, 2007 issue of *Environment and Urbanization* reports that 634 million people live in coastal areas within 30 feet (9.1 m) of sea level. The study also reported that about two thirds of the world's cities with over five million people are located in these low-lying coastal areas. Future sea level rise could lead to potentially catastrophic difficulties for shore-based communities in the next centuries: for example, many major cities such as Venice, London, New Orleans, and New York already need storm-surge defenses, and would need more if the sea level rose, though they also face issues such as subsidence. However, modest increases in sea level are likely to be offset when cities adapt by constructing sea walls or through relocating.

Re-insurance company Swiss Re estimates an economic loss for southeast Florida in 2030, of $33 billion from climate-related damages. Miami has been listed as "the number-one most vulnerable city worldwide" in terms of potential damage to property from storm-related flooding and sea-level rise.

Habitats

Coastal and Polar habitats are facing drastic changes as consequence of rising sea levels. Loss of ice in the Arctic may force local species to migrate in search of a new home. If seawater continues to approach inland, problems related to contaminated soils and flooded wetlands may occur. Also, fish, birds, and coastal plants could lose parts of their habitat. In 2016 it was reported that the Bramble Cay melomys, which lived on a Great Barrier Reef island, had probably become extinct because of sea level rises.

Extreme Sea Level Rise Events

Downturn of Atlantic meridional overturning circulation (AMOC), has been tied to extreme regional sea level rise (1-in-850 year event). Between 2009–2010, coastal sea levels north of New York City increased by 128 mm within two years. This jump is unprecedented in the tide gauge records, which collects data since a couple of centuries.

Sea Level Measurement

Satellites

Jason-1 continues the same sea surface measurements begun by TOPEX/Poseidon. It will be followed by the Ocean Surface Topography Mission on Jason-2 and by a planned future Jason-3

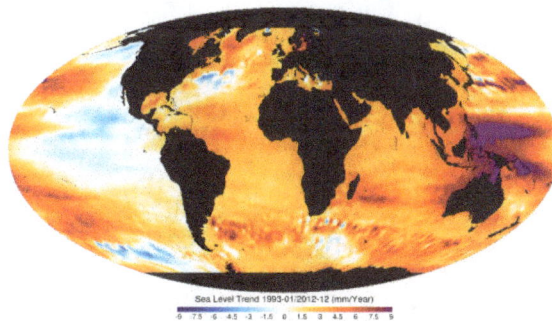

1993–2012 Sea level trends from satellite altimetry

In 1992 the TOPEX/Poseidon satellite was launched to record the change in sea level. Current rates of sea level rise from satellite altimetry have been estimated in the range of 2.9–3.4 ± 0.4–0.6 mm per year for 1993–2010. This exceeds those from tide gauges. It is unclear whether this represents an increase over the last decades; variability; true differences between satellites and tide gauges; or problems with satellite calibration. Due to calibration errors of the first satellite – Topex/Poseidon, sea levels have been slightly overestimated until 2015, which resulted in masking of ongoing sea level rise acceleration.

Tide Gauge

Amsterdam

The longest running sea-level measurements, NAP or Amsterdam Ordnance Datum established in 1675, are recorded in Amsterdam, the Netherlands. About 25 percent of the Netherlands lies beneath sea level, while more than 50 percent of this nation's area would be inundated by temporary floods if it did not have an extensive levee system.

Australia

In Australia, data collected by the Commonwealth Scientific and Industrial Research Organisation (CSIRO) show the current global mean sea level trend to be 3.2 mm/yr., a doubling of the rate of the total increase of about 210mm that was measured from 1880 to 2009, which reflected an average annual rise over the entire 129-year period of about 1.6 mm/year.

Australian record collection has a long time horizon, including measurements by an amateur meteorologist beginning in 1837 and measurements taken from a sea-level benchmark struck on a small cliff on the Isle of the Dead near the Port Arthur convict settlement on 1 July 1841. These records, when compared with data recorded by modern tide gauges, reinforce the recent comparisons of the historic sea level rise of about 1.6 mm/year, with the sharp acceleration in recent decades.

Continuing extensive sea level data collection by Australia's (CSIRO) is summarized in in its finding of mean sea level trend to be 3.2 mm/yr. As of 2003 the National Tidal Centre of the Bureau of Meteorology managed 32 tide gauges covering the entire Australian coastline, with some measurements available starting in 1880.

United States

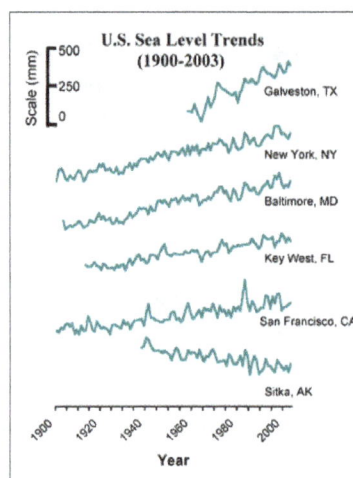

US sea-level trends 1900–2003

Tide gauges in the United States reveal considerable variation because some land areas are rising and some are sinking. For example, over the past 100 years, the rate of sea level rise varied from about an increase of 0.36 inches (9.1 mm) per year along the Louisiana Coast (due to land sinking), to a drop of a few inches per decade in parts of Alaska (due to post-glacial rebound). The rate of sea level rise increased during the 1993–2003 period compared with the longer-term average (1961–2003), although it is unclear whether the faster rate reflected a short-term variation or an increase in the long-term trend.

One study showed no acceleration in sea level rise in US tide gauge records during the 20th century. However, another study found that the rate of rise for the US Atlantic coast during the 20th century was far higher than during the previous two thousand years.

Adaptation

In 2008, the Dutch *Delta Commission* (Deltacommissie), advised in a report that the Netherlands would need a massive new building program to strengthen the country's water defenses against the anticipated effects of global warming for the next 190 years. This commission was created in September 2007, after the damage caused by Hurricane Katrina prompted reflection and preparations. Those included drawing up worst-case plans for evacuations. The plan included more than €100 billion (US$144 bn), in new spending through the year 2100 to take measures, such as broadening coastal dunes and strengthening sea and river dikes. The commission said the country must plan for a rise in the North Sea up to 1.3 metres (4 ft 3 in) by 2100, rather than the previously projected 0.80 metres (2 ft 7 in), and plan for a 2–4 metre (6.5–13 feet) rise by 2200.

The New York City Panel on Climate Change (NPCC), is an effort to prepare the New York City area for climate change.

Miami Beach is spending $500 million in the next years to address sea-level rise. Actions include a pump drainage system, and to raise roadways and sidewalks.

Geophysical Fluid Dynamics

Geophysical fluid dynamics is the study of naturally occurring, large-scale flows on Earth and other planets. It is applied to the motion of fluids in the ocean and outer core, and to gases in the atmosphere of Earth and other planets. Two features that are common to many of the phenomena studied in geophysical fluid dynamics are rotation of the fluid due to the planetary rotation and stratification (layering). The applications of geophysical fluid dynamics do not generally include the circulation of the mantle, which is the subject of geodynamics, or fluid phenomena in the magnetosphere. Smaller scale flow features (those negligibly influenced by the rotation of the Earth) are the province of fields such as hydrology, physical oceanography and meteorology.

Fundamentals

To describe the flow of geophysical fluids, equations are needed for conservation of momentum (or Newton's second law) and conservation of energy. The former leads to the Navier-Stokes equations. Further approximations are generally made. First, the fluid is assumed to be incompressible. Remarkably, this works well even for a highly compressible fluid like air as long as sound and shock waves can be ignored. Second, the fluid is assumed to be a Newtonian fluid, meaning that there is a linear relation between the shear stress τ and the strain u, for example

$$\tau = \mu \frac{du}{dx},$$

where μ is the viscosity. Under these assumptions the Navier-Stokes equations are

$$\rho\left(\overbrace{\underbrace{\frac{\partial \mathbf{v}}{\partial t}}_{\substack{\text{Eulerian} \\ \text{acceleration}}} + \underbrace{\mathbf{v} \cdot \nabla \mathbf{v}}_{\text{Advection}}}^{\text{Inertia (per volume)}} \right) = \overbrace{-\nabla p}^{\substack{\text{Pressure} \\ \text{gradient}}} + \underbrace{\mu \nabla^2 \mathbf{v}}_{\text{Viscosity}} + \underbrace{\mathbf{f}}_{\substack{\text{Other} \\ \text{body} \\ \text{forces}}} .$$

The left hand side represents the acceleration that a small parcel of fluid would experience in a reference frame that moved with the parcel (a Lagrangian frame of reference). In a stationary (Eulerian) frame of reference, this acceleration is divided into the local rate of change of velocity and advection, a measure of the rate of flow in or out of a small region.

The equation for energy conservation is essentially an equation for heat flow. If heat is transported by conduction, the heat flow is governed by a diffusion equation. If there are also buoyancy effects, for example hot air rising, then natural convection can occur.

Buoyancy and Stratification

Fluid that is less dense than its surroundings tends to rise until it has the same density as its surroundings. If there is not much energy input to the system, it will tend to become stratified. On a large scale, Earth's atmosphere is divided into a series of layers. Going upwards from the ground, these are the troposphere, stratosphere, mesosphere, thermosphere, and exosphere.

The density of air is mainly determined by temperature and water vapor content, the density of sea water by temperature and salinity, and the density of lake water by temperature. Where stratification occurs, there may be thin layers in which temperature or some other property changes more rapidly with height or depth than the surrounding fluid. Depending on the main sources of buoyancy, this layer may be called a pycnocline (density), thermocline (temperature), halocline (salinity), or chemocline (chemistry, including oxygenation).

The same buoyancy that gives rise to stratification also drives gravity waves. If the gravity waves occur within the fluid, they are called internal waves.

In modeling buoyancy-driven flows, the Navier-Stokes equations are modified using the Boussinesq approximation. This ignores variations in density except where they are multiplied by the gravitational acceleration g.

If the pressure depends only on density and vice versa, the fluid dynamics are called barotropic. In the atmosphere, this corresponds to a lack of fronts, as in the tropics. If there are fronts, the flow is baroclinic, and instabilities such as cyclones can occur.

Plate Tectonics

Plate tectonics is a scientific theory describing the large-scale motion of Earth's lithosphere. The theoretical model builds on the concept of continental drift developed during the first few decades of the 20th century. The geoscientific community accepted plate-tectonic theory after seafloor spreading was validated in the late 1950s and early 1960s.

The lithosphere, which is the rigid outermost shell of a planet (the crust and upper mantle), is broken up into tectonic plates. The Earth's lithosphere is composed of seven or eight major plates (depending on how they are defined) and many minor plates. Where the plates meet, their relative motion determines the type of boundary: convergent, divergent, or transform. Earthquakes, volcanic activity, mountain-building, and oceanic trench formation occur along these plate boundaries. The relative movement of the plates typically ranges from zero to 100 mm annually.

Tectonic plates are composed of oceanic lithosphere and thicker continental lithosphere, each topped by its own kind of crust. Along convergent boundaries, subduction carries plates into the mantle; the material lost is roughly balanced by the formation of new (oceanic) crust along divergent margins by seafloor spreading. In this way, the total surface of the lithosphere remains the same. This prediction of plate tectonics is also referred to as the conveyor belt principle. Earlier theories (that still have some supporters) propose gradual shrinking (contraction) or gradual expansion of the globe.

Tectonic plates are able to move because the Earth's lithosphere has greater strength than the underlying asthenosphere. Lateral density variations in the mantle result in convection. Plate movement is thought to be driven by a combination of the motion of the seafloor away from the spreading ridge (due to variations in topography and density of the crust, which result in differences in gravitational forces) and drag, with downward suction, at the subduction zones. Another explanation lies in the different forces generated by tidal forces of the Sun and Moon. The relative importance of each of these factors and their relationship to each other is unclear, and still the subject of much debate.

Key Principles

The outer layers of the Earth are divided into the lithosphere and asthenosphere. This is based on differences in mechanical properties and in the method for the transfer of heat. Mechanically, the lithosphere is cooler and more rigid, while the asthenosphere is hotter and flows more easily. In terms of heat transfer, the lithosphere loses heat by conduction, whereas the asthenosphere also transfers heat by convection and has a nearly adiabatic temperature gradient. This division should not be confused with the *chemical* subdivision of these same layers into the mantle (comprising both the asthenosphere and the mantle portion of the lithosphere) and the crust: a given piece of mantle may be part of the lithosphere or the asthenosphere at different times depending on its temperature and pressure.

The key principle of plate tectonics is that the lithosphere exists as separate and distinct *tectonic plates*, which ride on the fluid-like (visco-elastic solid) asthenosphere. Plate motions range up to a typical 10–40 mm/year (Mid-Atlantic Ridge; about as fast as fingernails grow), to about 160 mm/year (Nazca Plate; about as fast as hair grows). The driving mechanism behind this movement is described below.

Tectonic lithosphere plates consist of lithospheric mantle overlain by either or both of two types of crustal material: oceanic crust (in older texts called *sima* from silicon and magnesium) and continental crust (*sial* from silicon and aluminium). Average oceanic lithosphere is typically 100 km (62 mi) thick; its thickness is a function of its age: as time passes, it conductively cools and subjacent cooling mantle is added to its base. Because it is formed at mid-ocean ridges and spreads outwards, its thickness is therefore a function of its distance from the mid-ocean ridge where it was formed. For a typical distance that oceanic lithosphere must travel before being subducted, the thickness varies from about 6 km (4 mi) thick at mid-ocean ridges to greater than 100 km (62 mi) at subduction zones; for shorter or longer distances, the subduction zone (and therefore also the mean) thickness becomes smaller or larger, respectively. Continental lithosphere is typically ~200 km thick, though this varies considerably between basins, mountain ranges, and stable cratonic interiors of continents. The two types of crust also differ in thickness, with continental crust being considerably thicker than oceanic (35 km vs. 6 km).

The location where two plates meet is called a *plate boundary*. Plate boundaries are commonly associated with geological events such as earthquakes and the creation of topographic features such as mountains, volcanoes, mid-ocean ridges, and oceanic trenches. The majority of the world's active volcanoes occur along plate boundaries, with the Pacific Plate's Ring of Fire being the most active and widely known today. These boundaries are discussed in further detail below. Some volcanoes occur in the interiors of plates, and these have been variously attributed to internal plate deformation and to mantle plumes.

As explained above, tectonic plates may include continental crust or oceanic crust, and most plates contain both. For example, the African Plate includes the continent and parts of the floor of the Atlantic and Indian Oceans. The distinction between oceanic crust and continental crust is based on their modes of formation. Oceanic crust is formed at sea-floor spreading centers, and continental crust is formed through arc volcanism and accretion of terranes through tectonic processes, though some of these terranes may contain ophiolite sequences, which are pieces of oceanic crust considered to be part of the continent when they exit the standard cycle of formation and spreading centers and subduction beneath continents. Oceanic crust is also denser than continental crust owing to their different compositions. Oceanic crust is denser because it has less silicon and more heavier elements ("mafic") than continental crust ("felsic"). As a result of this density stratification, oceanic crust generally lies below sea level (for example most of the Pacific Plate), while continen-tal crust buoyantly projects above sea level.

Types of Plate Boundaries

Three types of plate boundaries exist, with a fourth, mixed type, characterized by the way the plates move relative to each other. They are associated with different types of surface phenomena. The different types of plate boundaries are:

1. *Transform boundaries (Conservative)* occur where two lithospheric plates slide, or perhaps more accurately, grind past each other along transform faults, where plates are neither created nor destroyed. The relative motion of the two plates is either sinistral (left side toward the observer) or dextral (right side toward the observer). Transform faults occur across a spreading center. Strong earthquakes can occur along a fault. The San Andreas Fault in California is an example of a transform boundary exhibiting dextral motion.

2. *Divergent boundaries (Constructive)* occur where two plates slide apart from each other. At zones of ocean-to-ocean rifting, divergent boundaries form by seafloor spreading, allowing for the formation of new ocean basin. As the continent splits, the ridge forms at the spreading center, the ocean basin expands, and finally, the plate area increases causing many small volcanoes and/or shallow earthquakes. At zones of continent-to-continent rifting, divergent boundaries may cause new ocean basin to form as the continent splits, spreads, the central rift collapses, and ocean fills the basin. Active zones of Mid-ocean ridges (e.g., Mid-Atlantic Ridge and East Pacific Rise), and continent-to-continent rifting (such as Africa's East African Rift and Valley, Red Sea) are examples of divergent boundaries.

3. *Convergent boundaries (Destructive)* (or *active margins*) occur where two plates slide toward each other to form either a subduction zone (one plate moving underneath the other) or a continental collision. At zones of ocean-to-continent subduction (e.g. the Andes mountain range in South America, and the Cascade Mountains in Western United States), the dense oceanic lithosphere plunges beneath the less dense continent. Earthquakes trace the path of the downward-moving plate as it descends into asthenosphere, a trench forms, and as the subducted plate is heated it releases volatiles, mostly water from hydrous minerals, into the surrounding mantle. The addition of water lowers the melting point of the mantle material above the subducting slab, causing it to melt. The magma that results typically leads to volcanism. At zones of ocean-to-ocean subduction (e.g. Aleutian islands, Mariana Islands, and the Japanese island arc), older, cooler, denser crust slips beneath less dense crust. This causes earthquakes and a deep trench to form in an arc shape. The upper mantle of the subducted plate then heats and magma rises to form curving chains of volcanic islands. Deep marine trenches are typically associated with subduction zones, and the basins that develop along the active boundary are often called "foreland basins". Closure of ocean basins can occur at continent-to-continent boundaries (e.g., Himalayas and Alps): collision between masses of granitic continental lithosphere; neither mass is subducted; plate edges are compressed, folded, uplifted.

4. *Plate boundary zones* occur where the effects of the interactions are unclear, and the boundaries, usually occurring along a broad belt, are not well defined and may show various types of movements in different episodes.

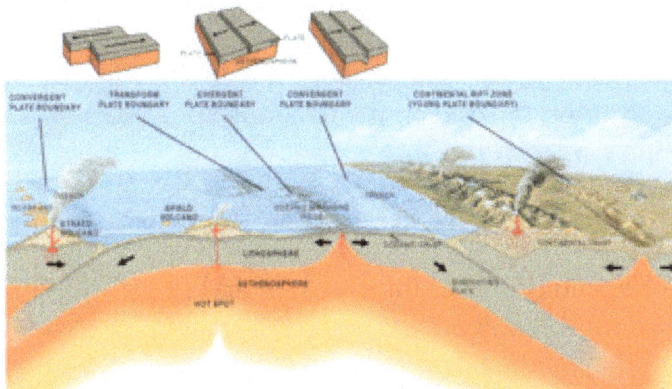

Three types of plate boundary.

Driving Forces of Plate Motion

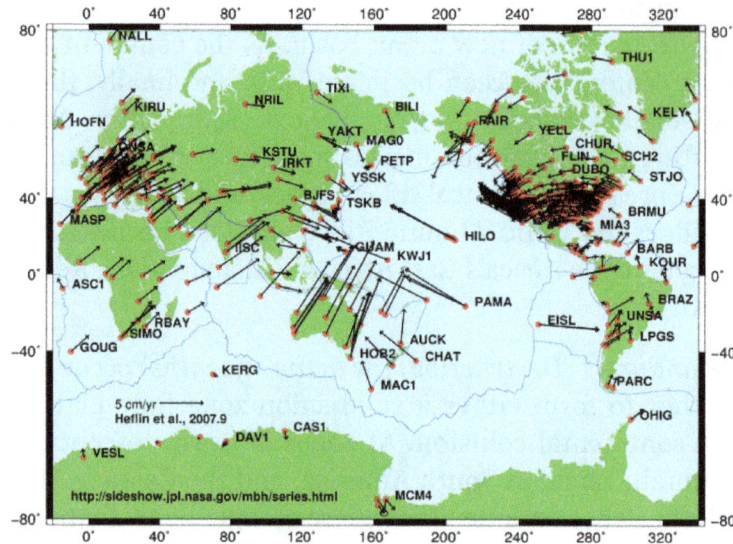

Plate motion based on Global Positioning System (GPS) satellite data from NASA JPL. The vectors show direction and magnitude of motion.

It is generally accepted that tectonic plates are able to move because of the relative density of oceanic lithosphere and the relative weakness of the asthenosphere. Dissipation of heat from the mantle is acknowledged to be the original source of the energy required to drive plate tectonics through convection or large scale upwelling and doming. The current view, though still a matter of some debate, asserts that as a consequence, a powerful source of plate motion is generated due to the excess density of the oceanic lithosphere sinking in subduction zones. When the new crust forms at mid-ocean ridges, this oceanic lithosphere is initially less dense than the underlying asthenosphere, but it becomes denser with age as it conductively cools and thickens. The greater density of old lithosphere relative to the underlying asthenosphere allows it to sink into the deep mantle at subduction zones, providing most of the driving force for plate movement. The weakness of the asthenosphere allows the tectonic plates to move easily towards a subduction zone. Although subduction is thought to be the strongest force driving plate motions, it cannot be the only force since there are plates such as the North American Plate which are moving, yet are nowhere being subducted. The same is true for the enormous Eurasian Plate. The sources of plate motion are a matter of intensive research and discussion among scientists. One of the main points is that the kinematic pattern of the movement itself should be separated clearly from the possible geodynamic mechanism that is invoked as the driving force of the observed movement, as some patterns may be explained by more than one mechanism. In short, the driving forces advocated at the moment can be divided into three categories based on the relationship to the movement: mantle dynamics related, gravity related (mostly secondary forces).

Driving Forces Related to Mantle Dynamics

For much of the last quarter century, the leading theory of the driving force behind tectonic plate motions envisaged large scale convection currents in the upper mantle which are transmitted through the asthenosphere. This theory was launched by Arthur Holmes and some forerunners in the 1930s and was immediately recognized as the solution for the acceptance of the theory as originally discussed in the papers of Alfred Wegener in the early years of the century. However, despite its acceptance, it was long debated in the scientific community because the leading ("fixist") theory still envisaged a static Earth without moving continents up until the major breakthroughs of the early sixties.

Two- and three-dimensional imaging of Earth's interior (seismic tomography) shows a varying lateral density distribution throughout the mantle. Such density variations can be material (from rock chemistry), mineral (from variations in mineral structures), or thermal (through thermal expansion and contraction from heat energy). The manifestation of this varying lateral density is mantle convection from buoyancy forces.

How mantle convection directly and indirectly relates to plate motion is a matter of ongoing study and discussion in geodynamics. Somehow, this energy must be transferred to the lithosphere for tectonic plates to move. There are essentially two types of forces that are thought to influence plate motion: friction and gravity.

- Basal drag (friction): Plate motion driven by friction between the convection currents in the asthenosphere and the more rigid overlying lithosphere.

- Slab suction (gravity): Plate motion driven by local convection currents that exert a downward pull on plates in subduction zones at ocean trenches. Slab suction may occur in a geodynamic setting where basal tractions continue to act on the plate as it dives into the mantle (although perhaps to a greater extent acting on both the under and upper side of the slab).

Lately, the convection theory has been much debated as modern techniques based on 3D seismic tomography still fail to recognize these predicted large scale convection cells. Therefore, alternative views have been proposed:

In the theory of plume tectonics developed during the 1990s, a modified concept of mantle convection currents is used. It asserts that super plumes rise from the deeper mantle and are the drivers or substitutes of the major convection cells. These ideas, which find their roots in the early 1930s with the so-called "fixistic" ideas of the European and Russian Earth Science Schools, find resonance in the modern theories which envisage hot spots/mantle plumes which remain fixed and are overridden by oceanic and continental lithosphere plates over time and leave their traces in the geological record (though these phenomena are not invoked as real driving mechanisms, but rather as modulators). Modern theories that continue building on the older mantle doming concepts and see plate movements as a secondary phenomena are beyond the scope of this page and are discussed elsewhere (for example on the plume tectonics page).

Another theory is that the mantle flows neither in cells nor large plumes but rather as a series of channels just below the Earth's crust, which then provide basal friction to the lithosphere. This theory, called "surge tectonics", became quite popular in geophysics and geodynamics during the 1980s and 1990s.

Driving Forces Related to Gravity

Forces related to gravity are usually invoked as secondary phenomena within the framework of a more general driving mechanism such as the various forms of mantle dynamics described above.

Gravitational sliding away from a spreading ridge: According to many authors, plate motion is driven by the higher elevation of plates at ocean ridges. As oceanic lithosphere is formed at spreading ridges from hot mantle material, it gradually cools and thickens with age (and thus adds distance from the ridge). Cool oceanic lithosphere is significantly denser than the hot mantle material from which it is derived and so with increasing thickness it gradually subsides into the mantle to compensate the greater load. The result is a slight lateral incline with increased distance from the ridge axis.

This force is regarded as a secondary force and is often referred to as "ridge push". This is a misnomer as nothing is "pushing" horizontally and tensional features are dominant along ridges. It is more accurate to refer to this mechanism as gravitational sliding as variable topography across the totality of the plate can vary considerably and the topography of spreading ridges is only the most prominent feature. Other mechanisms generating this gravitational secondary force include flexural bulging of the lithosphere before it dives underneath an adjacent plate which produces a clear topographical feature that can offset, or at least affect, the influence of topographical ocean ridges, and mantle plumes and hot spots, which are postulated to impinge on the underside of tectonic plates.

Slab-pull: Current scientific opinion is that the asthenosphere is insufficiently competent or rigid to directly cause motion by friction along the base of the lithosphere. Slab pull is therefore most widely thought to be the greatest force acting on the plates. In this current understanding, plate motion is mostly driven by the weight of cold, dense plates sinking into the mantle at trenches. Recent models indicate that trench suction plays an important role as well. However, as the North American Plate is nowhere being subducted, yet it is in motion presents a problem. The same holds for the African, Eurasian, and Antarctic plates.

Gravitational sliding away from mantle doming: According to older theories, one of the driving mechanisms of the plates is the existence of large scale asthenosphere/mantle domes which cause the gravitational sliding of lithosphere plates away from them. This gravitational sliding represents a secondary phenomenon of this basically vertically oriented mechanism. This can act on various scales, from the small scale of one island arc up to the larger scale of an entire ocean basin.

Driving Forces Related to Earth Rotation

Alfred Wegener, being a meteorologist, had proposed tidal forces and pole flight force as the main driving mechanisms behind continental drift; however, these forces were considered far too small to cause continental motion as the concept then was of continents plowing through oceanic crust. Therefore, Wegener later changed his position and asserted that convection currents are the main driving force of plate tectonics in the last edition of his book in 1929.

However, in the plate tectonics context (accepted since the seafloor spreading proposals of Heezen, Hess, Dietz, Morley, Vine, and Matthews during the early 1960s), oceanic crust is suggested to be in motion *with* the continents which caused the proposals related to Earth rotation to be reconsidered. In more recent literature, these driving forces are:

1. Tidal drag due to the gravitational force the Moon (and the Sun) exerts on the crust of the Earth

2. Global deformation of the geoid due to small displacements of rotational pole with respect to the Earth's crust;

3. Other smaller deformation effects of the crust due to wobbles and spin movements of the Earth rotation on a smaller time scale.

Forces that are small and generally negligible are:

1. The Coriolis force

2. The centrifugal force, which is treated as a slight modification of gravity.

For these mechanisms to be overall valid, systematic relationships should exist all over the globe between the orientation and kinematics of deformation and the geographical latitudinal and longitudinal grid of the Earth itself. Ironically, these systematic relations studies in the second half of the nineteenth century and the first half of the twentieth century underline exactly the opposite: that the plates had not moved in time, that the deformation grid was fixed with respect to the Earth equator and axis, and that gravitational driving forces were generally acting vertically and caused only local horizontal movements (the so-called pre-plate tectonic, "fixist theories"). Later studies (discussed below on this page), therefore, invoked many of the relationships recognized during this pre-plate tectonics period to support their theories.

Of the many forces discussed in this paragraph, tidal force is still highly debated and defended as a possible principle driving force of plate tectonics. The other forces are only used in global geodynamic models not using plate tectonics concepts (therefore beyond the discussions treated in this section) or proposed as minor modulations within the overall plate tectonics model.

In 1973, George W. Moore of the USGS and R. C. Bostrom presented evidence for a general westward drift of the Earth's lithosphere with respect to the mantle. He concluded that tidal forces (the tidal lag or "friction") caused by the Earth's rotation and the forces acting upon it by the Moon are a driving force for plate tectonics. As the Earth spins eastward beneath the moon, the moon's gravity ever so slightly pulls the Earth's surface layer back westward, just as proposed by Alfred Wegener. In a more recent 2006 study, scientists reviewed and advocated these earlier proposed ideas. It has also been suggested recently in Lovett (2006) that this observation may also explain why Venus and Mars have no plate tectonics, as Venus has no moon and Mars' moons are too small to have significant tidal effects on the planet. In a recent paper, it was suggested that, on the other hand, it can easily be observed that many plates are moving north and eastward, and that the dominantly westward motion of the Pacific ocean basins derives simply from the eastward bias of the Pacific spreading center (which is not a predicted manifestation of such lunar forces). In the same paper the authors admit, however, that relative to the lower mantle, there is a slight westward component in the motions of all the plates. They demonstrated though that the westward drift, seen only for the past 30 Ma, is attributed to the increased dominance of the steadily growing and accelerating Pacific plate. The debate is still open.

Relative Significance of Each Driving Force Mechanism

The vector of a plate's motion is a function of all the forces acting on the plate; however, therein lies the problem regarding the degree to which each process contributes to the overall motion of each tectonic plate.

The diversity of geodynamic settings and the properties of each plate result from the impact of the various processes actively driving each individual plate. One method of dealing with this problem is to consider the relative rate at which each plate is moving as well as the evidence related to the significance of each process to the overall driving force on the plate.

One of the most significant correlations discovered to date is that lithospheric plates attached to downgoing (subducting) plates move much faster than plates not attached to subducting plates. The Pacific plate, for instance, is essentially surrounded by zones of subduction (the so-called Ring of Fire) and moves much faster than the plates of the Atlantic basin, which are attached (perhaps one could say 'welded') to adjacent continents instead of subducting plates. It is thus thought that forces associated with the downgoing plate (slab pull and slab suction) are the driving forces which determine the motion of plates, except for those plates which are not being subducted. The driving forces of plate motion continue to be active subjects of on-going research within geophysics and tectonophysics.

Development of the Theory

Summary

Detailed map showing the tectonic plates with their movement vectors.

In line with other previous and contemporaneous proposals, in 1912 the meteorologist Alfred Wegener amply described what he called continental drift, expanded in his 1915 book *The Origin of Continents and Oceans* and the scientific debate started that would end up fifty years later in the theory of plate tectonics. Starting from the idea (also expressed by his forerunners) that the present continents once formed a single land mass (which was called Pangea later on) that drifted apart, thus releasing the continents from the Earth's mantle and likening them to "icebergs" of low density granite floating on a sea of denser basalt. Supporting evidence for the idea came from the dove-tailing outlines of South America's east coast and Africa's west coast, and from the matching of the rock formations along these edges. Confirmation of their previous contiguous nature also came from the fossil plants *Glossopteris* and *Gangamopteris*, and the therapsid or mammal-like reptile *Lystrosaurus*, all widely distributed over South America, Africa, Antarctica, India and Australia. The evidence for such an erstwhile joining of these continents was patent to field geologists working in the southern hemisphere. The South African Alex du Toit put together a mass of such information in his 1937 publication *Our Wan-*

dering Continents, and went further than Wegener in recognising the strong links between the Gondwana fragments.

But without detailed evidence and a force sufficient to drive the movement, the theory was not generally accepted: the Earth might have a solid crust and mantle and a liquid core, but there seemed to be no way that portions of the crust could move around. Distinguished scientists, such as Harold Jeffreys and Charles Schuchert, were outspoken critics of continental drift.

Despite much opposition, the view of continental drift gained support and a lively debate started between "drifters" or "mobilists" (proponents of the theory) and "fixists" (opponents). During the 1920s, 1930s and 1940s, the former reached important milestones proposing that convection currents might have driven the plate movements, and that spreading may have occurred below the sea within the oceanic crust. Concepts close to the elements now incorporated in plate tectonics were proposed by geophysicists and geologists (both fixists and mobilists) like Vening-Meinesz, Holmes, and Umbgrove.

One of the first pieces of geophysical evidence that was used to support the movement of lithospheric plates came from paleomagnetism. This is based on the fact that rocks of different ages show a variable magnetic field direction, evidenced by studies since the mid–nineteenth century. The magnetic north and south poles reverse through time, and, especially important in paleotectonic studies, the relative position of the magnetic north pole varies through time. Initially, during the first half of the twentieth century, the latter phenomenon was explained by introducing what was called "polar wander", i.e., it was assumed that the north pole location had been shifting through time. An alternative explanation, though, was that the continents had moved (shifted and rotated) relative to the north pole, and each continent, in fact, shows its own "polar wander path". During the late 1950s it was successfully shown on two occasions that these data could show the validity of continental drift: by Keith Runcorn in a paper in 1956, and by Warren Carey in a symposium held in March 1956.

The second piece of evidence in support of continental drift came during the late 1950s and early 60s from data on the bathymetry of the deep ocean floors and the nature of the oceanic crust such as magnetic properties and, more generally, with the development of marine geology which gave evidence for the association of seafloor spreading along the mid-oceanic ridges and magnetic field reversals, published between 1959 and 1963 by Heezen, Dietz, Hess, Mason, Vine & Matthews, and Morley.

Simultaneous advances in early seismic imaging techniques in and around Wadati-Benioff zones along the trenches bounding many continental margins, together with many other geophysical (e.g. gravimetric) and geological observations, showed how the oceanic crust could disappear into the mantle, providing the mechanism to balance the extension of the ocean basins with shortening along its margins.

All this evidence, both from the ocean floor and from the continental margins, made it clear around 1965 that continental drift was feasible and the theory of plate tectonics, which was defined in a series of papers between 1965 and 1967, was born, with all its extraordinary explanatory and predictive power. The theory revolutionized the Earth sciences, explaining a diverse range of geological phenomena and their implications in other studies such as paleogeography and paleobiology.

Continental Drift

In the late 19th and early 20th centuries, geologists assumed that the Earth's major features were fixed, and that most geologic features such as basin development and mountain ranges could be explained by vertical crustal movement, described in what is called the geosynclinal theory. Generally, this was placed in the context of a contracting planet Earth due to heat loss in the course of a relatively short geological time.

Alfred Wegener in Greenland in the winter of 1912-13.

It was observed as early as 1596 that the opposite coasts of the Atlantic Ocean—or, more precisely, the edges of the continental shelves—have similar shapes and seem to have once fitted together.

Since that time many theories were proposed to explain this apparent complementarity, but the assumption of a solid Earth made these various proposals difficult to accept.

The discovery of radioactivity and its associated heating properties in 1895 prompted a re-examination of the apparent age of the Earth. This had previously been estimated by its cooling rate and assumption the Earth's surface radiated like a black body. Those calculations had implied that, even if it started at red heat, the Earth would have dropped to its present temperature in a few tens of millions of years. Armed with the knowledge of a new heat source, scientists realized that the Earth would be much older, and that its core was still sufficiently hot to be liquid.

By 1915, after having published a first article in 1912, Alfred Wegener was making serious arguments for the idea of continental drift in the first edition of *The Origin of Continents and Oceans*. In that book (re-issued in four successive editions up to the final one in 1936), he noted how the east coast of South America and the west coast of Africa looked as if they were once attached. Wegener was not the first to note this (Abraham Ortelius, Antonio Snider-Pellegrini, Eduard Suess, Roberto Mantovani and Frank Bursley Taylor preceded him just to mention a few), but he was the first to marshal significant fossil and paleo-topographical and climatological evidence to support this simple observation (and was supported in this by researchers such as Alex du Toit). Furthermore, when the rock strata of the margins of separate continents are very similar it suggests that these rocks were formed in the same way, implying that they were joined initially. For instance, parts of Scotland and Ireland contain rocks very similar to those found in Newfoundland and New Brunswick. Furthermore, the Caledonian Mountains of Europe and parts of the Appalachian Mountains of North America are very similar in structure and lithology.

However, his ideas were not taken seriously by many geologists, who pointed out that there was no apparent mechanism for continental drift. Specifically, they did not see how continental rock could plow through the much denser rock that makes up oceanic crust. Wegener could not explain the force that drove continental drift, and his vindication did not come until after his death in 1930.

Floating Continents, Paleomagnetism, and Seismicity Zones

Preliminary Determination of Epicenters
358,214 Events, 1963 - 1998

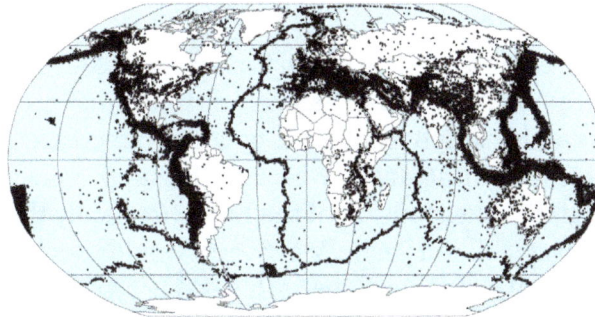

Global earthquake epicenters, 1963–1998

As it was observed early that although granite existed on continents, seafloor seemed to be composed of denser basalt, the prevailing concept during the first half of the twentieth century was that there were two types of crust, named "sial" (continental type crust) and "sima" (oceanic type crust). Furthermore, it was supposed that a static shell of strata was present under the continents. It therefore looked apparent that a layer of basalt (sial) underlies the continental rocks.

However, based on abnormalities in plumb line deflection by the Andes in Peru, Pierre Bouguer had deduced that less-dense mountains must have a downward projection into the denser layer underneath. The concept that mountains had "roots" was confirmed by George B. Airy a hundred years later, during study of Himalayan gravitation, and seismic studies detected corresponding density variations. Therefore, by the mid-1950s, the question remained unresolved as to whether mountain roots were clenched in surrounding basalt or were floating on it like an iceberg.

During the 20th century, improvements in and greater use of seismic instruments such as seismographs enabled scientists to learn that earthquakes tend to be concentrated in specific areas, most notably along the oceanic trenches and spreading ridges. By the late 1920s, seismologists were beginning to identify several prominent earthquake zones parallel to the trenches that typically were inclined 40–60° from the horizontal and extended several hundred kilometers into the Earth. These zones later became known as Wadati-Benioff zones, or simply Benioff zones, in honor of the seismologists who first recognized them, Kiyoo Wadati of Japan and Hugo Benioff of the United States. The study of global seismicity greatly advanced in the 1960s with the establishment of the Worldwide Standardized Seismograph Network (WWSSN) to monitor the compliance of the 1963 treaty banning above-ground testing of nuclear weapons. The much improved data from the

WWSSN instruments allowed seismologists to map precisely the zones of earthquake concentration worldwide.

Meanwhile, debates developed around the phenomena of polar wander. Since the early debates of continental drift, scientists had discussed and used evidence that polar drift had occurred because continents seemed to have moved through different climatic zones during the past. Furthermore, paleomagnetic data had shown that the magnetic pole had also shifted during time. Reasoning in an opposite way, the continents might have shifted and rotated, while the pole remained relatively fixed. The first time the evidence of magnetic polar wander was used to support the movements of continents was in a paper by Keith Runcorn in 1956, and successive papers by him and his students Ted Irving (who was actually the first to be convinced of the fact that paleomagnetism supported continental drift) and Ken Creer.

This was immediately followed by a symposium in Tasmania in March 1956. In this symposium, the evidence was used in the theory of an expansion of the global crust. In this hypothesis the shifting of the continents can be simply explained by a large increase in size of the Earth since its formation. However, this was unsatisfactory because its supporters could offer no convincing mechanism to produce a significant expansion of the Earth. Certainly there is no evidence that the moon has expanded in the past 3 billion years; other work would soon show that the evidence was equally in support of continental drift on a globe with a stable radius.

During the thirties up to the late fifties, works by Vening-Meinesz, Holmes, Umbgrove, and numerous others outlined concepts that were close or nearly identical to modern plate tectonics theory. In particular, the English geologist Arthur Holmes proposed in 1920 that plate junctions might lie beneath the sea, and in 1928 that convection currents within the mantle might be the driving force. Often, these contributions are forgotten because:

- At the time, continental drift was not accepted.

- Some of these ideas were discussed in the context of abandoned fixistic ideas of a deforming globe without continental drift or an expanding Earth.

- They were published during an episode of extreme political and economic instability that hampered scientific communication.

- Many were published by European scientists and at first not mentioned or given little credit in the papers on sea floor spreading published by the American researchers in the 1960s.

Mid-oceanic Ridge Spreading and Convection

In 1947, a team of scientists led by Maurice Ewing utilizing the Woods Hole Oceanographic Institution's research vessel *Atlantis* and an array of instruments, confirmed the existence of a rise in the central Atlantic Ocean, and found that the floor of the seabed beneath the layer of sediments consisted of basalt, not the granite which is the main constituent of continents. They also found that the oceanic crust was much thinner than continental crust. All these new findings raised important and intriguing questions.

The new data that had been collected on the ocean basins also showed particular characteristics regarding the bathymetry. One of the major outcomes of these datasets was that all along the globe, a system of mid-oceanic ridges was detected. An important conclusion was that along this system, new ocean floor was being created, which led to the concept of the "Great Global Rift". This was described in the crucial paper of Bruce Heezen (1960), which would trigger a real revolution in thinking. A profound consequence of seafloor spreading is that new crust was, and still is, being continually created along the oceanic ridges. Therefore, Heezen advocated the so-called "expanding Earth" hypothesis of S. Warren Carey. So, still the question remained: how can new crust be continuously added along the oceanic ridges without increasing the size of the Earth? In reality, this question had been solved already by numerous scientists during the forties and the fifties, like Arthur Holmes, Vening-Meinesz, Coates and many others: The crust in excess disappeared along what were called the oceanic trenches, where so-called "subduction" occurred. Therefore, when various scientists during the early sixties started to reason on the data at their disposal regarding the ocean floor, the pieces of the theory quickly fell into place.

The question particularly intrigued Harry Hammond Hess, a Princeton University geologist and a Naval Reserve Rear Admiral, and Robert S. Dietz, a scientist with the U.S. Coast and Geodetic Survey who first coined the term *seafloor spreading*. Dietz and Hess (the former published the same idea one year earlier in *Nature*, but priority belongs to Hess who had already distributed an unpublished manuscript of his 1962 article by 1960) were among the small handful who really understood the broad implications of sea floor spreading and how it would eventually agree with the, at that time, unconventional and unaccepted ideas of continental drift and the elegant and mobilistic models proposed by previous workers like Holmes.

In the same year, Robert R. Coats of the U.S. Geological Survey described the main features of island arc subduction in the Aleutian Islands. His paper, though little noted (and even ridiculed) at the time, has since been called "seminal" and "prescient". In reality, it actually shows that the work by the European scientists on island arcs and mountain belts performed and published during the 1930s up until the 1950s was applied and appreciated also in the United States.

If the Earth's crust was expanding along the oceanic ridges, Hess and Dietz reasoned like Holmes and others before them, it must be shrinking elsewhere. Hess followed Heezen, suggesting that new oceanic crust continuously spreads away from the ridges in a conveyor belt–like motion. And, using the mobilistic concepts developed before, he correctly concluded that many millions of years later, the oceanic crust eventually descends along the continental margins where oceanic trenches – very deep, narrow canyons – are formed, e.g. along the rim of the Pacific Ocean basin. The important step Hess made was that convection currents would be the driving force in this process, arriving at the same conclusions as Holmes had decades before with the only difference that the thinning of the ocean crust was performed using Heezen's mechanism of spreading along the ridges. Hess therefore concluded that the Atlantic Ocean was expanding while the Pacific Ocean was shrinking. As old oceanic crust is "consumed" in the trenches (like Holmes and others, he thought this was done by thickening of the continental lithosphere, not, as now understood, by underthrusting at a larger scale of the oceanic crust itself into the mantle), new magma rises and erupts along the spreading ridges to form new crust. In effect, the ocean basins are perpetually being "recycled," with the creation of new crust and the destruction of old oceanic lithosphere occurring simultaneously. Thus, the new mobilistic concepts neatly explained why the Earth does not get bigger with sea floor spreading, why there is so little sediment accumulation on the ocean floor, and why oceanic rocks are much younger than continental rocks.

Magnetic Striping

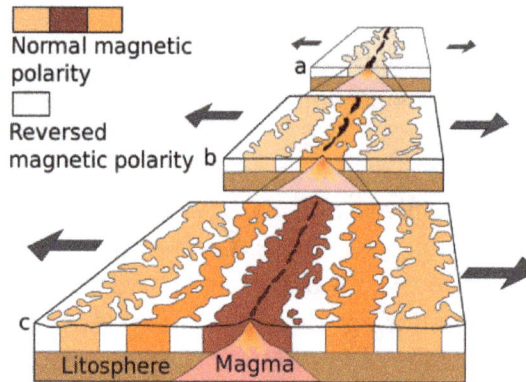

Seafloor magnetic striping.

Beginning in the 1950s, scientists like Victor Vacquier, using magnetic instruments (magnetometers) adapted from airborne devices developed during World War II to detect submarines, began recognizing odd magnetic variations across the ocean floor. This finding, though unexpected, was not entirely surprising because it was known that basalt—the iron-rich, volcanic rock making up the ocean floor—contains a strongly magnetic mineral (magnetite) and can locally distort compass readings. This distortion was recognized by Icelandic mariners as early as the late 18th century. More important, because the presence of magnetite gives the basalt measurable magnetic properties, these newly discovered magnetic variations provided another means to study the deep ocean floor. When newly formed rock cools, such magnetic materials recorded the Earth's magnetic field at the time.

A demonstration of magnetic striping. (The darker the color is, the closer it is to normal polarity)

As more and more of the seafloor was mapped during the 1950s, the magnetic variations turned out not to be random or isolated occurrences, but instead revealed recognizable patterns. When these magnetic patterns were mapped over a wide region, the ocean floor showed a zebra-like pattern: one stripe with normal polarity and the adjoining stripe with reversed polarity. The overall pattern, defined by these alternating bands of normally and reversely polarized rock, became known as magnetic striping, and was published by Ron G. Mason and co-workers in 1961, who did not find, though, an explanation for these data in terms of sea floor spreading, like Vine, Matthews and Morley a few years later.

The discovery of magnetic striping called for an explanation. In the early 1960s scientists such as Heezen, Hess and Dietz had begun to theorise that mid-ocean ridges mark structurally weak zones where the ocean floor was being ripped in two lengthwise along the ridge crest. New magma from deep within the Earth rises easily through these weak zones and eventually erupts along the crest of the ridges to create new oceanic crust. This process, at first denominated the "conveyer belt hypothesis" and later called seafloor spreading, operating over many millions of years continues to form new ocean floor all across the 50,000 km-long system of mid-ocean ridges.

Only four years after the maps with the "zebra pattern" of magnetic stripes were published, the link between sea floor spreading and these patterns was correctly placed, independently by Lawrence Morley, and by Fred Vine and Drummond Matthews, in 1963, now called the Vine-Matthews-Morley hypothesis. This hypothesis linked these patterns to geomagnetic reversals and was supported by several lines of evidence:

1. the stripes are symmetrical around the crests of the mid-ocean ridges; at or near the crest of the ridge, the rocks are very young, and they become progressively older away from the ridge crest;

2. the youngest rocks at the ridge crest always have present-day (normal) polarity;

3. stripes of rock parallel to the ridge crest alternate in magnetic polarity (normal-reversed-normal, etc.), suggesting that they were formed during different epochs documenting the (already known from independent studies) normal and reversal episodes of the Earth's magnetic field.

By explaining both the zebra-like magnetic striping and the construction of the mid-ocean ridge system, the seafloor spreading hypothesis (SFS) quickly gained converts and represented another major advance in the development of the plate-tectonics theory. Furthermore, the oceanic crust now came to be appreciated as a natural "tape recording" of the history of the geomagnetic field reversals (GMFR) of the Earth's magnetic field. Today, extensive studies are dedicated to the calibration of the normal-reversal patterns in the oceanic crust on one hand and known timescales derived from the dating of basalt layers in sedimentary sequences (magnetostratigraphy) on the other, to arrive at estimates of past spreading rates and plate reconstructions.

Definition and Refining of the Theory

After all these considerations, Plate Tectonics (or, as it was initially called "New Global Tectonics") became quickly accepted in the scientific world, and numerous papers followed that defined the concepts:

- In 1965, Tuzo Wilson who had been a promotor of the sea floor spreading hypothesis and continental drift from the very beginning added the concept of transform faults to the model, completing the classes of fault types necessary to make the mobility of the plates on the globe work out.

- A symposium on continental drift was held at the Royal Society of London in 1965 which must be regarded as the official start of the acceptance of plate tectonics by the scientific community, and which abstracts are issued as Blacket, Bullard & Runcorn (1965). In this symposium, Edward Bullard and co-workers showed with a computer calculation how the continents along both sides of the Atlantic would best fit to close the ocean, which became known as the famous "Bullard's Fit".

- In 1966 Wilson published the paper that referred to previous plate tectonic reconstructions, introducing the concept of what is now known as the "Wilson Cycle".

- In 1967, at the American Geophysical Union's meeting, W. Jason Morgan proposed that the Earth's surface consists of 12 rigid plates that move relative to each other.

- Two months later, Xavier Le Pichon published a complete model based on 6 major plates with their relative motions, which marked the final acceptance by the scientific community of plate tectonics.

- In the same year, McKenzie and Parker independently presented a model similar to Morgan's using translations and rotations on a sphere to define the plate motions.

Implications for Biogeography

Continental drift theory helps biogeographers to explain the disjunct biogeographic distribution of present-day life found on different continents but having similar ancestors. In particular, it explains the Gondwanan distribution of ratites and the Antarctic flora.

Plate Reconstruction

Reconstruction is used to establish past (and future) plate configurations, helping determine the shape and make-up of ancient supercontinents and providing a basis for paleogeography.

Defining Plate Boundaries

Current plate boundaries are defined by their seismicity. Past plate boundaries within existing plates are identified from a variety of evidence, such as the presence of ophiolites that are indicative of vanished oceans.

Past Plate Motions

Tectonic Motion First Began Around Three Billion Years Ago.

Various types of quantitative and semi-quantitative information are available to constrain past plate motions. The geometric fit between continents, such as between west Africa and South America is still an important part of plate reconstruction. Magnetic stripe patterns provide a reliable guide to relative plate motions going back into the Jurassic period. The tracks of hotspots give absolute reconstructions, but these are only available back to the Cretaceous. Older reconstructions rely mainly on paleomagnetic pole data, although these only constrain the latitude and rotation, but not the longitude. Combining poles of different ages in a particular plate to produce apparent polar wander paths provides a method for comparing the motions of different plates through time. Additional evidence comes from the distribution of certain sedimentary rock types, faunal provinces shown by particular fossil groups, and the position of orogenic belts.

Formation and Break-up of Continents

The movement of plates has caused the formation and break-up of continents over time, including occasional formation of a supercontinent that contains most or all of the continents. The supercontinent Columbia or Nuna formed during a period of 2,000 to 1,800 million years ago and broke up about 1,500 to 1,300 million years ago. The supercontinent Rodinia is thought to have formed about 1 billion years ago and to have embodied most or all of Earth's continents, and broken up into eight continents around 600 million years ago. The eight continents later re-assembled into another supercontinent called Pangaea; Pangaea broke up into Laurasia (which became North America and Eurasia) and Gondwana (which became the remaining continents).

The Himalayas, the world's tallest mountain range, are assumed to have been formed by the collision of two major plates. Before uplift, they were covered by the Tethys Ocean.

Current Plates

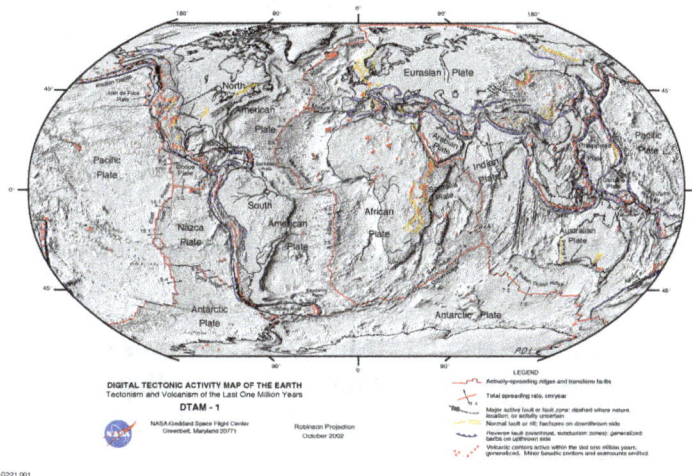

DIGITAL TECTONIC ACTIVITY MAP OF THE EARTH
Tectonism and Volcanism of the Last One Million Years
DTAM - 1

Depending on how they are defined, there are usually seven or eight "major" plates: African, Antarctic, Eurasian, North American, South American, Pacific, and Indo-Australian. The latter is sometimes subdivided into the Indian and Australian plates.

There are dozens of smaller plates, the seven largest of which are the Arabian, Caribbean, Juan de Fuca, Cocos, Nazca, Philippine Sea and Scotia.

The current motion of the tectonic plates is today determined by remote sensing satellite data sets, calibrated with ground station measurements.

Other Celestial Bodies (Planets, Moons)

The appearance of plate tectonics on terrestrial planets is related to planetary mass, with more massive planets than Earth expected to exhibit plate tectonics. Earth may be a borderline case, owing its tectonic activity to abundant water (silica and water form a deep eutectic.)

Venus

Venus shows no evidence of active plate tectonics. There is debatable evidence of active tectonics in the planet's distant past; however, events taking place since then (such as the plausible and generally accepted hypothesis that the Venusian lithosphere has thickened greatly over the course of several hundred million years) has made constraining the course of its geologic record difficult. However, the numerous well-preserved impact craters have been utilized as a dating method to approximately date the Venusian surface (since there are thus far no known samples of Venusian rock to be dated by more reliable methods). Dates derived are dominantly in the range 500 to 750 million years ago, although ages of up to 1,200 million years ago have been calculated. This research has led to the fairly well accepted hypothesis that Venus has undergone an essentially complete volcanic resurfacing at least once in its distant past, with the last event taking place approximately within the range of estimated surface ages. While the mechanism of such an impressive thermal event remains a debated issue in Venusian geosciences, some scientists are advocates of processes involving plate motion to some extent.

One explanation for Venus's lack of plate tectonics is that on Venus temperatures are too high for significant water to be present. The Earth's crust is soaked with water, and water plays an important role in the development of shear zones. Plate tectonics requires weak surfaces in the crust along which crustal slices can move, and it may well be that such weakening never took place on Venus because of the absence of water. However, some researchers[who?] remain convinced that plate tectonics is or was once active on this planet.

Mars

Mars is considerably smaller than Earth and Venus, and there is evidence for ice on its surface and in its crust.

In the 1990s, it was proposed that Martian Crustal Dichotomy was created by plate tectonic processes. Scientists today disagree, and think that it was created either by upwelling within the Martian mantle that thickened the crust of the Southern Highlands and formed Tharsis or by a giant impact that excavated the Northern Lowlands.

Valles Marineris may be a tectonic boundary.

Observations made of the magnetic field of Mars by the *Mars Global Surveyor* spacecraft in 1999 showed patterns of magnetic striping discovered on this planet. Some scientists interpreted these as requiring plate tectonic processes, such as seafloor spreading. However, their data fail a "magnetic reversal test", which is used to see if they were formed by flipping polarities of a global magnetic field.

Icy Satellites

Some of the satellites of Jupiter have features that may be related to plate-tectonic style deformation, although the materials and specific mechanisms may be different from plate-tectonic activity on Earth. On 8 September 2014, NASA reported finding evidence of plate tectonics on Europa, a satellite of Jupiter—the first sign of such geological activity on another world other than Earth.

Titan, the largest moon of Saturn, was reported to show tectonic activity in images taken by the *Huygens* probe, which landed on Titan on January 14, 2005.

Exoplanets

On Earth-sized planets, plate tectonics is more likely if there are oceans of water; however, in 2007, two independent teams of researchers came to opposing conclusions about the likelihood of plate tectonics on larger super-earths with one team saying that plate tectonics would be episodic or stagnant and the other team saying that plate tectonics is very likely on super-earths even if the planet is dry.

Ocean Turbidity

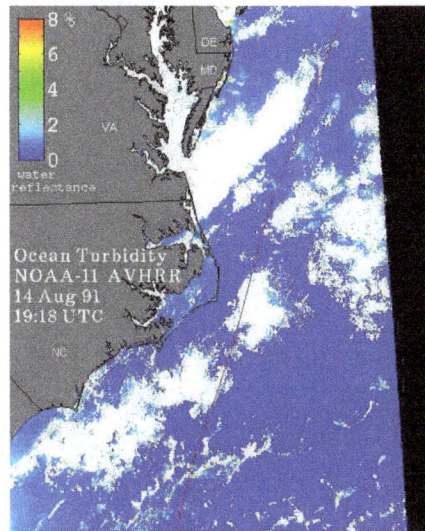

Visualisation of the Ocean Turbidity of the ocean just before Hurricane Bob (August 14, 1991)

Ocean turbidity is a measure of the amount of cloudiness or haziness in sea water caused by individual particles that are too small to be seen without magnification. Highly turbid ocean waters are those with a large number of scattering particulates in them. In both highly absorbing and highly scattering waters, visibility into the water is reduced. The highly scattering (turbid) water still reflects a lot of light while the highly absorbing water, such as a blackwater river or lake, is very dark. The scattering particles that cause the water to be turbid can be composed of many things, including sediments and phytoplankton.

Measurement

There are a number of ways to measure ocean turbidity, including autonomous remote vehicles, shipcasts and satellites.

From a satellite, a proxy measurement of the water turbidity can be made by examining the amount of reflectance in the visible region of the electromagnetic spectrum. For the Advanced Very High Resolution Radiometer (AVHRR), the logical choice is band 1, covering wavelengths 580 to 680 nanometres, the orange and red. In order to make derived products that are comparable over time and space, an atmospheric correction is required. To do this, the effects of Rayleigh scattering are calculated based on the satellite viewing angle and the solar zenith angle and then subtracted from the band 1 radiance. For an aerosol correction, band 2 in the near infrared is used. It is first corrected for Rayleigh scattering and then subtracted from the Rayleigh corrected band 1. The Rayleigh corrected band 2 is assumed to be aerosol radiance because no return signal from water in the near infrared is expected since water is highly absorbing at those wavelengths. Because bands 1 and 2 are relatively close on the electromagnetic spectrum, we can reasonably assume their aerosol radiances are the same.

In these images the turbidity is quantified as the percent reflected light emerging from the water column in a range of 0 to 8 percent. The reflectance percentage can be correlated to attenuation, Secchi disk depth or total suspended solids although the exact relationship will vary regionally and depends on the optical properties of the water. For example, in Florida Bay, 10% reflectance corresponds to a sediment concentration of 30 milligram/litre and a Secchi depth of 0.5 metre. These relationships are approximately linear so that 5% reflectance would correspond to a sediment concentration of approximately 15 milligram/litre and a Secchi depth of 1 metre. In the Mississippi River plume regions these same reflectance values would represent sediment concentrations that are about ten times or more higher.

Hurricanes

As one would expect, the majority of these images reveal large increases in turbidity in the regions where a hurricane has made landfall. The increases are primarily due to sediments that have been resuspended from the shallow bottom regions. In areas near shore some of the signal may also be due to sediments eroded from beaches as well as from sediment laden river plumes. In some cases a post-hurricane phytoplankton bloom due to increased nutrient availability may perhaps be detectable.

The examination of the turbidity after the passing of a hurricane can have potentially many uses for coastal resource management including:

- identifying regional "hot spots" where the erosion could be expected to be most severe

- estimating the total sediment concentration that has been mobilized by the hurricane

- determining the spatial extent of the sediment mobilization

- identifying the extent and contribution of river plumes

- assessing and predicting potential ecosystem impacts

With regard to these uses, determining the regions of high turbidity will allow managers to best decide on response strategies as well as help ensure that post-hurricane resources are most effectively utilized.

Interpreting Images

Only a small fraction of the light incident on the ocean will be reflected and received by the satellite. The probability for a photon to reflect and exit the ocean decreases exponentially with length of its path through the water because the ocean is an absorbing medium. The more ocean a photon must travel through, the greater its chances of being absorbed by something. After absorption, it will eventually become part of the ocean's heat reservoir. The absorption and scattering character-istics of a water body determine the rate of vertical light attenuation and set a limit to the depths contributing to a satellite signal. A reasonable rule of thumb is that 90 percent of the signal coming from the water that is seen by the satellite is from the first attenuation length. How deep this is depends on the absorption and scattering properties of both the water itself and other constituents in the water. For wavelengths in the near infrared and longer, the penetration depth varies from a metre to a few micrometres. For band 1, the penetration depth will usually be between 1 and 10 metres. If the water has a large turbidity spike below 10 metres, the spike is unlikely to be seen by a satellite.

For very shallow clear water there is a good chance the bottom may be seen. For example, in the Bahamas, the water is quite clear and only a few metres deep, resulting in an apparent high turbid-ity because the bottom reflects a lot of the band 1 light. For areas with consistently high turbidity signals, particularly areas with relatively clear water, part of the signal may be due to bottom re-flection. Normally this will not be a problem with a post-hurricane turbidity image since the storm easily resuspends enough sediment such that bottom reflection is negligible.

Clouds are also problematic for the interpretation of satellite derived turbidity. Cloud removal al-gorithms perform a satisfactory job for pixels that are fully cloudy. Partially cloudy pixels are much harder to identify and typically result in false high turbidity estimates. High turbidity values near clouds are suspect.

References

- Richard Stenger (September 19, 2000). "Flotilla of sensors to monitor world's oceans". CNN. Archived from the original on 6 November 2007. Retrieved 2007-10-28.

- "Argo Floats : How do we measure the ocean?" (Youtube video). Integrated Marine Observing Strategy. 10 March 2014. Retrieved 8 October 2014.

- Davidson, Helen (30 January 2014). "Scientists to launch bio robots in Indian Ocean to study its 'interior biology'". The Guardian. Retrieved 8 October 2014.

- Kenneth Chang (June 24, 1999). "An Ear to Ocean Temperature". ABC News. Archived from the original on 2003-10-06. Retrieved 2007-10-23.

- Potter, J. R. (1994). "ATOC: Sound Policy or Enviro-Vandalism? Aspects of a Modern Media-Fueled Policy Issue". 3. The Journal of Environment & Development. pp. 47–62. doi:10.1177/107049659400300205. Retrieved 2009-11-20.

- National Research Council (2000). Marine mammals and low-frequency sound: Progress since 1994. Washington, D.C.: National Academy Press.

- Frankel, A. S.; C. W. Clark (2000). "Behavioral responses of humpback whales (Megaptera novaeangliae) to full-scale ATOC signals". 108. Journal of the Acoustical Society of America. pp. 1–8. Retrieved 2009-08-15.

- Frankel, A. S.; C. W. Clark (2002). "ATOC and other factors affecting distribution and abundance of humpback whales (Megaptera novaeangliae) off the north shore of Kauai". 18. Marine Mammal Science. pp. 664–662. Retrieved 2009-08-15.

- Mobley, J. R. (2005). "Assessing responses of humpback whales to North Pacific Acoustic Laboratory (NPAL) transmissions: Results of 2001-2003 aerial surveys north of Kauai". 117. Journal of the Acoustical Society of America. pp. 1666–1673. Retrieved 2009-08-15.

- Bombosch, A. (2014). "Predictive habitat modelling of humpback (Megaptera novaeangliae) and Antarctic minke (Balaenoptera bonaerensis) whales in the Southern Ocean as a planning tool for seismic surveys".

Deep-Sea Research Part I: Oceanographic Research Papers. Deep-Sea Research Part I. 91: 101–114. Bibcode:2014DSRI...91..101B. doi:10.1016/j.dsr.2014.05.017. Retrieved 2015-01-25.

- National Research Council (2003). Ocean Noise and Marine Mammals. National Academies Press. ISBN 978-0-309-08536-6. Retrieved 2015-01-25.

- Garrison, Tom (2009). Oceanography: An Invitation to Marine Science (7th ed.). Cengage Learning. p. 582. ISBN 9780495391937.

- Kump, Lee R.; Kasting, James F.; Crane, Robert G. (2003). The Earth System (2nd ed.). Upper Saddle River: Prentice Hall. pp. 162–164. ISBN 0-613-91814-2.

- Bows, Kevin; Bows, Alice (2011). "Beyond 'dangerous' climate change: emission scenarios for a new world" (PDF). Philosophical Transactions of the Royal Society A. 369 (1934): 20–44. Bibcode:2011RSPTA.369...20A. doi:10.1098/rsta.2010.0290. Retrieved 2011-05-22.

Ocean Observations and its Model

Ocean reanalysis is a method of combining historical ocean observations with a general ocean model. In recent time, a number of efforts have been put in to estimate the physical state of the ocean. This chapter provides the reader with a detailed insight on ocean observations and on the reanalysis of oceans.

Ocean Observations

The following are considered essential ocean climate variables by the Ocean Observations Panel for Climate (OOPC) that are currently feasible with current observational systems .

Ocean Climate Variables

Atmosphere Surface

Air Temperature

Precipitation (meteorology)

evaporation

Air Pressure, sea level pressure (SLP)

Surface radiative fluxes

Surface thermodynamic fluxes

Wind speed and direction

Surface wind stress

Water vapor

Ocean Surface

Sea surface temperature (SST)

Sea surface salinity (SSS)

Sea level

Sea state

Sea ice

Ocean current

Ocean color (for biological activity)

Carbon dioxide partial pressure (pCO2)

Ocean Subsurface

- Backscatter
- Carbon Dioxide
- Chlorophyll
- Conductivity
- Density
- Iron
- Irradiance
- Nutrients
  - Nitrate
- Methane
- Ocean current
  - Single Point
  - Water Column
- Ocean tracers
- Oxygen
- Phytoplankton
- Salinity
- Sigma-T
- Sound Velocity
- Temperature
- Turbidity

Ocean Observation Sources

Satellite

There is a composite network of satellites that generate observations. These include:

Type	Variables observed	Responsible organizations
Infrared (IR)	SST, sea ice	CEOS, IGOS, CGMS
AMSR-class microwave	SST, wind speed, sea ice	CEOS, IGOS, CGMS
Surface vector wind (two wide-swath scatterometers desired)	surface vector wind, sea ice	CEOS, IGOS, CGMS
Ocean color	chlorophyll concentration (biomass of phytoplankton)	IOCCG
high-precision altimetry	sea-level anomaly from steady state	CEOS, IGOS, CGMS
low-precision altimetry	sea level	CEOS, IGOS, CGMS
Synthetic aperture radar	sea ice, sea state	CEOS, IGOS, CGMS

In Situ

There is a composite network of in situ observations. These include:

Type	Variables observed	Responsible organizations
Global surface drifting buoy array with 5 degree resolution (1250 total)	SST, SLP, Current (based on position change)	JCOMM Data Buoy Cooperation Panel (DBCP)
Global tropical moored buoy network (about 120 moorings)	typically SST and surface vector wind, but can also include SLP, current, air-sea flux variables	JCOMM DBCP Tropical Moored Buoy Implementation Panel (TIP)
Volunteer Observing Ship (VOS) fleet	all feasible surface ECVs	JCOMM Ship Observations Team (SOT)
VOSClim	all feasible surface ECVs plus extensive ship metadata	JCOMM Ship Observations Team (SOT)
Global referencing mooring network (29 moorings)	all feasible surface ECVs	OceanSITES
GLOSS core sea-level network, plus regional/national networks	sea level	JCOMM GLOSS
Carbon VOS	pCO2, SST, SSS	IOCCP
Sea ice buoys	sea ice	JCOMM DBCP IABP and IPAB

Subsurface

There is a composite network of subsurface observations. These include:

Type	Variables observed	Responsible organizations
Repeat XBT (Expendable bathythermograph) line network (41 lines)	Temperature	JCOMM Ship Observations Team (SOT)

Global tropical moored buoy network (~120 moorings)	Temperature, salinity, current, other feasible autonomously observable ECVs	JCOMM DBCP Tropical Moored Buoy Implementation Panel (TIP)
Reference mooring network (29 moorings)	all autonomously observable ECVs	OceanSITES
Sustained and repeated ship-based hydrography network	All feasible ECVs, including those that depend on obtaining water samples	IOCCP, CLIVAR, other national efforts
Argo (oceanography) network	temperature, salinity, current	Argo
Critical current and transport monitoring	temperature, heat, freshwater, carbon transports, mass	CLIVAR, IOCCP, OceanSITES
Regional and global synthesis programmes	inferred currents, transports gridded fields of all ECVs	GODAE, CLIVAR, other national efforts
Cabled ocean observatories	audio, backscatter, chlorophyll, CO_2, conductivity, currents, density, Eh, gravity, iron, irradiance, methane, nitrate, oxygen, pressure, salinity, seismic, sigma-T, sound velocity, temperature, turbidity, video	Ocean Networks Canada, Monterey Accelerated Research System, Ocean Observatories Initiative, ALOHA, ESONET (European Seas Observatory NETwork), Dense Oceanfloor Network System for Earthquakes and Tsunamis (DONET), Fixed-Point Open Ocean Observatories (FixO3).

Accuracy of Measurements

The quality of *in situ* measurements is non-uniform across space, time and platforms. Different platforms employ a large variety of sensors, which operate in a wide range of often hostile environments and use different measurement protocols. Occasionally, buoys are left unattended for extended periods of time, while ships may involve a certain amount of the human-related impacts in data collection and transmission. Therefore, quality control is necessary before in situ data can be further used in scientific research or other applications. This is an example of quality control and monitoring of sea surface temperatures measured by ships and buoys, the iQuam system developed at NOAA/NESDIS/STAR, where statistics show the quality of *in situ* measurements of sea surface temperatures.

One of the problems facing real-time ocean observatories is the ability to provide a fast and accurate assessment of the data quality. Ocean Networks Canada is in the process of implementing real-time quality control on incoming data. For scalar data, the aim is to meet the guidelines of the Quality Assurance of Real Time Oceanographic Data (QARTOD) group. QARTOD is a US organization tasked with identifying issues involved with incoming real-time data from the U.S Integrated Ocean Observing System (IOOS). A large portion of their agenda is to create guidelines for how the quality of real-time data is to be determined and reported to the scientific community. Real-time data quality testing at Ocean Networks Canada includes tests designed to catch instrument failures and major spikes or data dropouts before the data is made available to the user. Real-time quality tests include meeting instrument manufacturer's standards and overall observatory/site ranges determined from previous data. Due to the positioning of some instrument platforms in highly productive areas, we have also designed dual-sensor tests e.g. for some conductivity sensors. The quality control testing is split into 3 separate categories. The first category is in real-time and tests the data before the data are parsed into the database. The second category is delayed-mode testing where archived data are subject to testing after a certain period of time. The third category is manual quality control by an Ocean Networks Canada data expert.

Historical Data Available

OceanSITES manages a set of links to various sources of available ocean data, including: the Hawaiian Ocean Timeseries (HOT), the JAMSTEC Kuroshio Extension Observatory (JKEO), Line W monitoring the North Atlantic's deep western boundary current, and others.

This site includes links to the ARGO Float Data, The Data Library and Archives (DLA), the Falmouth Monthly Climate Reports, Martha's Vineyard Coastal Observatory, the Multibeam Archive, the Seafloor Data and Observation Visualization Environment (SeaDOVE): A Web-served GIS Database of Multi-scalar Seafloor Data, Seafloor Sediments Data Collection, the Upper Ocean Mooring Data Archive, the U.S. GLOBEC Data System, U.S. JGOFS Data System, and the WHOI Ship Data-Grabber System.

There are a variety of data sets in a data library listed at Columbia University:

This library includes:

- LEVITUS94 is the World Ocean Atlas as of 1994, an atlas of objectively analyzed fields of major ocean parameters at the annual, seasonal, and monthly time scales. It is superseded by WOA98.

- NOAA NODC WOA98 is the World Ocean Atlas as of 1998, an atlas of objectively analyzed fields of major ocean parameters at monthly, seasonal, and annual time scales. Superseded by WOA01.

- NOAA NODC WOA01 is the World Ocean Atlas 2001, an atlas of objectively analyzed fields of major ocean parameters at monthly, seasonal, and annual time scales. Replaced by WOA05.

- NOAA NODC WOA05 is the World Ocean Atlas 2005, an atlas of objectively analyzed fields of major ocean parameters at monthly, seasonal, and annual time scales.

In situ observations spanning from the early 1700s to present are available from the International Comprehensive Ocean Atmosphere Data Set (ICOADS).

This data set includes observations of a number of the surface ocean and atmospheric variables from ships, moored and drifting buoys and C-MAN stations.

In 2006, Ocean Networks Canada began collecting high-resolution in-situ measurements from the seafloor in Saanich Inlet, near Victoria, British Columbia, Canada. Monitoring sites were later extended to the Strait of Georgia and 5 locations off the West coast of Vancouver Island, British Columbia, Canada. All historical measurements are freely available via Ocean Networks Canada's data portal, Oceans 2.0.

Ocean Reanalysis

Ocean reanalysis is a method of combining historical ocean observations with a general ocean model (typically a computational model) driven by historical estimates of surface winds, heat, and freshwater, by way of a data assimilation algorithm to reconstruct historical changes in the state of the ocean.

Historical observations are sparse and insufficient for understanding the history of the ocean and its circulation. By utilizing data assimilation techniques in combination with advanced computational models of the global ocean, researchers are able to interpolate the historical observations to all points in the ocean. This process has an analog in the construction of atmospheric reanalysis and is closely related to ocean state estimation.

Current Projects

A number of efforts have been initiated in recent years to apply data assimilation to estimate the physical state of the ocean, including temperature, salinity, currents, and sea level, in recent years. There are three alternative state estimation approaches. The first approach is used by the 'no-model' analyses, for which temperature or salinity observations update a first guess provided by climatological monthly estimates.

The second approach is that of the sequential data assimilation analyses, which move forward in time from a previous analysis using a numerical simulation of the evolving temperature and other variables produced by an ocean general circulation model. The simulation provides the first guess of the state of the ocean at the next analysis time, while corrections are made to this first guess based on observations of variables such as temperature, salinity, or sea level.

The third approach is 4D-Var, which in the implementation described uses the initial conditions and surface forcing as control variables to be modified in order to be consistent with the observations as well as a numerical representation of the equations of motion through iterative solution of a giant optimization problem.

Methodologies

No-model Approach

ISHII and LEVITUS begin with a first guess of the climatological monthly upper-ocean temperature based on climatologies produced by the NOAA National Oceanographic Data Center. The innovations are mapped onto the analysis levels. ISHII uses and alternative 3DVAR approach to do an objective mapping with a smaller decorrelation scale in midlatitudes (300 km) that elongates in the zonal direction by a factor of 3 at equatorial latitudes. LEVITUS begins similarly to ISHII, but uses the technique of Cressman and Barnes with a homogeneous scale of 555 km to objectively map the temperature innovation onto a uniform grid.

Sequential Approaches

The sequential approaches can be further divided into those using Optimal Interpolation and its more sophisticated cousin the Kalman Filter, and those using 3D-Var. Among those mentioned above, INGV and SODA use versions of Optimal Interpolation. CERFACS, GODAS, and GFDL all use 3DVar. "To date we are unaware of any attempt to use Kalman Filter for multi-decadal ocean reanalyses." The 4-Dimensional Local Ensemble Transform Kalman Filter (4D-LETKF) has been applied to the Geophysical Fluid Dynamics Laboratory's (GFDL) Modular Ocean Model (MOM2) for a 7-year ocean reanalysis from January 1997 – 2004.

Variational (4D-Var) Approach

One innovative attempt by GECCO has been made to apply 4D-Var to the decadal ocean estimation problem. This approach faces daunting computational challenges, but provides some interesting benefits including satisfying some conservation laws and the construction of the ocean model adjoint.

References

- "The Physical Oceanography Component of Hawaii Ocean Timeseries (HOT/PO)". Soest.hawaii.edu. Retrieved 14 January 2015.

- Jenkyns, Reyna (20 September 2010). "NEPTUNE Canada: Data integrity from the seafloor to your (Virtual) Door". Oceans 2010. doi:10.1109/OCEANS.2010.5664290. Retrieved 3 November 2015.

- Hunt, B.R., Kostelich E.J., Szunyogh, I. Efficient Data Assimilation for Spatiotemporal Chaos: A Local Ensemble Transform Kalman Filter. arXiv:physics/0511236 v1 28 Nov 2005. Dated May 24, 2006.

7

Impact of Pollution on Ocean

The impact of pollution on oceans has drastically increased. Eighty percent of marine pollution comes from land. Ocean population is as important as land pollution, and measures need to be taken to prevent it. The chapter serves as a source to understand marine pollution, and helps the reader to develop a deep understanding on the subject.

Marine Pollution

Marine pollution occurs when harmful, or potentially harmful, effects result from the entry into the ocean of chemicals, particles, industrial, agricultural and residential waste, noise, or the spread of invasive organisms. Eighty percent of marine pollution comes from land. Air pollution is also a contributing factor by carrying off pesticides or dirt into the ocean. Land and air pollution have proven to be harmful to marine life and its habitats.

The pollution often comes from non point sources such as agricultural runoff, wind-blown debris and dust. Nutrient pollution, a form of water pollution, refers to contamination by excessive inputs of nutrients. It is a primary cause of eutrophication of surface waters, in which excess nutrients, usually nitrogen or phosphorus, stimulate algae growth.

Many potentially toxic chemicals adhere to tiny particles which are then taken up by plankton and benthos animals, most of which are either deposit or filter feeders. In this way, the toxins are concentrated upward within ocean food chains. Many particles combine chemically in a manner highly depletive of oxygen, causing estuaries to become anoxic.

When pesticides are incorporated into the marine ecosystem, they quickly become absorbed into marine food webs. Once in the food webs, these pesticides can cause mutations, as well as diseases, which can be harmful to humans as well as the entire food web.

Toxic metals can also be introduced into marine food webs. These can cause a change to tissue matter, biochemistry, behaviour, reproduction, and suppress growth in marine life. Also, many animal feeds have a high fish meal or fish hydrolysate content. In this way, marine toxins can be transferred to land animals, and appear later in meat and dairy products.

History

Although marine pollution has a long history, significant international laws to counter it were only enacted in the twentieth century. Marine pollution was a concern during several United Nations Conferences on the Law of the Sea beginning in the 1950s. Most scientists believed that the oceans were so vast that they had unlimited ability to dilute, and thus render pollution harmless.

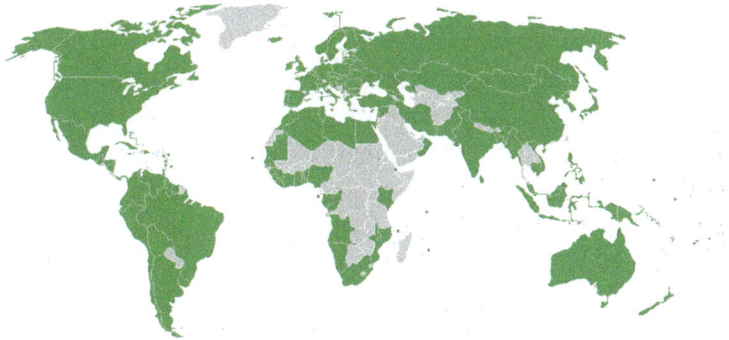

Parties to the MARPOL 73/78 convention on marine pollution

In the late 1950s and early 1960s, there were several controversies about dumping radioactive waste off the coasts of the United States by companies licensed by the Atomic Energy Commission, into the Irish Sea from the British reprocessing facility at Windscale, and into the Mediterranean Sea by the French Commissariat à l'Energie Atomique. After the Mediterranean Sea controversy, for example, Jacques Cousteau became a worldwide figure in the campaign to stop marine pollution. Marine pollution made further international headlines after the 1967 crash of the oil tanker Torrey Canyon, and after the 1969 Santa Barbara oil spill off the coast of California.

Marine pollution was a major area of discussion during the 1972 United Nations Conference on the Human Environment, held in Stockholm. That year also saw the signing of the Convention on the Prevention of Marine Pollution by Dumping of Wastes and Other Matter, sometimes called the London Convention. The London Convention did not ban marine pollution, but it established black and gray lists for substances to be banned (black) or regulated by national authorities (gray). Cyanide and high-level radioactive waste, for example, were put on the black list. The London Convention applied only to waste dumped from ships, and thus did nothing to regulate waste discharged as liquids from pipelines.

Pathways of Pollution

Septic river.

There are many different ways to categorize, and examine the inputs of pollution into our marine ecosystems. Patin (n.d.) notes that generally there are three main types of inputs of pollution into the ocean: direct discharge of waste into the oceans, runoff into the waters due to rain, and pollutants that are released from the atmosphere.

One common path of entry by contaminants to the sea are rivers. The evaporation of water from oceans exceeds precipitation. The balance is restored by rain over the continents entering rivers and then being returned to the sea. The Hudson in New York State and the Raritan in New Jersey, which empty at the northern and southern ends of Staten Island, are a source of mercury contamination of zooplankton (copepods) in the open ocean. The highest concentration in the filter-feeding copepods is not at the mouths of these rivers but 70 miles south, nearer Atlantic City, because water flows close to the coast. It takes a few days before toxins are taken up by the plankton.

Pollution is often classed as point source or nonpoint source pollution. Point source pollution occurs when there is a single, identifiable, and localized source of the pollution. An example is directly discharging sewage and industrial waste into the ocean. Pollution such as this occurs particularly in developing nations. Nonpoint source pollution occurs when the pollution comes from ill-defined and diffuse sources. These can be difficult to regulate. Agricultural runoff and wind blown debris are prime examples.

Direct Discharge

Acid mine drainage in the Rio Tinto River.

Pollutants enter rivers and the sea directly from urban sewerage and industrial waste discharges, sometimes in the form of hazardous and toxic wastes.

Inland mining for copper, gold. etc., is another source of marine pollution. Most of the pollution is simply soil, which ends up in rivers flowing to the sea. However, some minerals discharged in the course of the mining can cause problems, such as copper, a common industrial pollutant, which can interfere with the life history and development of coral polyps. Mining has a poor environmental track record. For example, according to the United States Environmental Protection Agency, mining has contaminated portions of the headwaters of over 40% of watersheds in the western continental US. Much of this pollution finishes up in the sea.

Land Runoff

Surface runoff from farming, as well as urban runoff and runoff from the construction of roads, buildings, ports, channels, and harbours, can carry soil and particles laden with carbon, nitrogen, phosphorus, and minerals. This nutrient-rich water can cause fleshy algae and phytoplankton to thrive in coastal areas; known as algal blooms, which have the potential to create hypoxic conditions by using all available oxygen.

Polluted runoff from roads and highways can be a significant source of water pollution in coastal areas. About 75% of the toxic chemicals that flow into Puget Sound are carried by stormwater that runs off paved roads and driveways, rooftops, yards and other developed land.

Ship Pollution

A cargo ship pumps ballast water over the side.

Ships can pollute waterways and oceans in many ways. Oil spills can have devastating effects. While being toxic to marine life, polycyclic aromatic hydrocarbons (PAHs), found in crude oil, are very difficult to clean up, and last for years in the sediment and marine environment.

Oil spills are probably the most emotive of marine pollution events. However, while a tanker wreck may result in extensive newspaper headlines, much of the oil in the world's seas comes from other smaller sources, such as tankers discharging ballast water from oil tanks used on return ships, leaking pipelines or engine oil disposed of down sewers.

Discharge of cargo residues from bulk carriers can pollute ports, waterways and oceans. In many instances vessels intentionally discharge illegal wastes despite foreign and domestic regulation prohibiting such actions. It has been estimated that container ships lose over 10,000 containers at sea each year (usually during storms). Ships also create noise pollution that disturbs natural wildlife, and water from ballast tanks can spread harmful algae and other invasive species.

Ballast water taken up at sea and released in port is a major source of unwanted exotic marine life. The invasive freshwater zebra mussels, native to the Black, Caspian and Azov seas, were probably transported to the Great Lakes via ballast water from a transoceanic vessel. Meinesz believes that one of the worst cases of a single invasive species causing harm to an ecosystem can be attributed to a seemingly harmless jellyfish. *Mnemiopsis leidyi*, a species of comb jellyfish that spread so it now inhabits estuaries in many parts of the world. It was first introduced in 1982, and thought to have been transported to the Black Sea in a ship's ballast water. The population of the jellyfish shot up exponentially and, by 1988, it was wreaking havoc upon the local fishing industry. "The anchovy catch fell from 204,000 tons in 1984 to 200 tons in 1993; sprat from 24,600 tons in 1984 to 12,000 tons in 1993; horse mackerel from 4,000 tons in 1984 to zero in 1993." Now that the jellyfish have exhausted the zooplankton, including fish larvae, their numbers have fallen dramatically, yet they continue to maintain a stranglehold on the ecosystem.

Invasive species can take over once occupied areas, facilitate the spread of new diseases, introduce new genetic material, alter underwater seascapes and jeopardize the ability of native species to obtain food. Invasive species are responsible for about $138 billion annually in lost revenue and management costs in the US alone.

Atmospheric Pollution

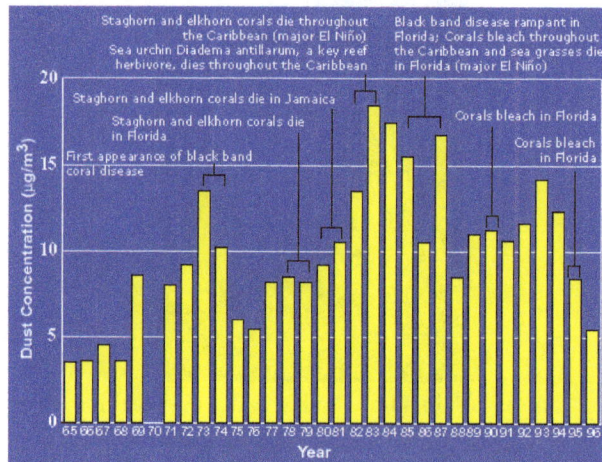

Graph linking atmospheric dust to various coral deaths across the Caribbean Sea and Florida

Another pathway of pollution occurs through the atmosphere. Wind blown dust and debris, including plastic bags, are blown seaward from landfills and other areas. Dust from the Sahara moving around the southern periphery of the subtropical ridge moves into the Caribbean and Florida during the warm season as the ridge builds and moves northward through the subtropical Atlantic. Dust can also be attributed to a global transport from the Gobi and Taklamakan deserts across Korea, Japan, and the Northern Pacific to the Hawaiian Islands. Since 1970, dust outbreaks have worsened due to periods of drought in Africa. There is a large variability in dust transport to the Caribbean and Florida from year to year; however, the flux is greater during positive phases of the North Atlantic Oscillation. The USGS links dust events to a decline in the health of coral reefs across the Caribbean and Florida, primarily since the 1970s.

Climate change is raising ocean temperatures and raising levels of carbon dioxide in the atmosphere. These rising levels of carbon dioxide are acidifying the oceans. This, in turn, is altering aquatic ecosystems and modifying fish distributions, with impacts on the sustainability of fisheries and the livelihoods of the communities that depend on them. Healthy ocean ecosystems are also important for the mitigation of climate change.

Deep Sea Mining

Deep sea mining is a relatively new mineral retrieval process that takes place on the ocean floor. Ocean mining sites are usually around large areas of polymetallic nodules or active and extinct hydrothermal vents at about 1,400 – 3,700 meters below the ocean's surface. The vents create sulfide deposits, which contain precious metals such as silver, gold, copper, manganese, cobalt, and zinc. The deposits are mined using either hydraulic pumps or bucket systems that take ore to the surface to be processed. As with all mining operations, deep sea mining raises questions about environmental damages to the surrounding areas

Because deep sea mining is a relatively new field, the complete consequences of full scale mining operations are unknown. However, experts are certain that removal of parts of the sea floor will result in disturbances to the benthic layer, increased toxicity of the water column and sediment plumes from tailings. Removing parts of the sea floor disturbs the habitat of benthic organisms, possibly, depending on the type of mining and location, causing permanent disturbances. Aside from direct impact of mining the area, leakage, spills and corrosion would alter the mining area's chemical makeup.

Among the impacts of deep sea mining, sediment plumes could have the greatest impact. Plumes are caused when the tailings from mining (usually fine particles) are dumped back into the ocean, creating a cloud of particles floating in the water. Two types of plumes occur: near bottom plumes and surface plumes. Near bottom plumes occur when the tailings are pumped back down to the mining site. The floating particles increase the turbidity, or cloudiness, of the water, clogging filter-feeding apparatuses used by benthic organisms. Surface plumes cause a more serious problem. Depending on the size of the particles and water currents the plumes could spread over vast areas. The plumes could impact zooplankton and light penetration, in turn affecting the food web of the area.

Types of Pollution

Acidification

Island with fringing reef in the Maldives. Coral reefs are dying around the world.

The oceans are normally a natural carbon sink, absorbing carbon dioxide from the atmosphere. Because the levels of atmospheric carbon dioxide are increasing, the oceans are becoming more acidic. The potential consequences of ocean acidification are not fully understood, but there are concerns that structures made of calcium carbonate may become vulnerable to dissolution, affecting corals and the ability of shellfish to form shells.

Oceans and coastal ecosystems play an important role in the global carbon cycle and have removed about 25% of the carbon dioxide emitted by human activities between 2000 and 2007 and about half the anthropogenic CO_2 released since the start of the industrial revolution. Rising ocean temperatures and ocean acidification means that the capacity of the ocean carbon sink will gradually get weaker, giving rise to global concerns expressed in the Monaco and Manado Declarations.

A report from NOAA scientists published in the journal Science in May 2008 found that large amounts of relatively acidified water are upwelling to within four miles of the Pacific continental shelf area of North America. This area is a critical zone where most local marine life lives or is born. While the paper dealt only with areas from Vancouver to northern California, other continental shelf areas may be experiencing similar effects.

A related issue is the methane clathrate reservoirs found under sediments on the ocean floors. These trap large amounts of the greenhouse gas methane, which ocean warming has the potential to release. In 2004 the global inventory of ocean methane clathrates was estimated to occupy between one and five million cubic kilometres. If all these clathrates were to be spread uniformly across the ocean floor, this would translate to a thickness between three and fourteen metres. This estimate corresponds to 500–2500 gigatonnes carbon (Gt C), and can be compared with the 5000 Gt C estimated for all other fossil fuel reserves.

Eutrophication

Polluted lagoon.

Effect of eutrophication on marine benthic life

Eutrophication is an increase in chemical nutrients, typically compounds containing nitrogen or phosphorus, in an ecosystem. It can result in an increase in the ecosystem's primary productivity (excessive plant growth and decay), and further effects including lack of oxygen and severe reductions in water quality, fish, and other animal populations.

The biggest culprit are rivers that empty into the ocean, and with it the many chemicals used as fertilizers in agriculture as well as waste from livestock and humans. An excess of oxygen depleting chemicals in the water can lead to hypoxia and the creation of a dead zone.

Estuaries tend to be naturally eutrophic because land-derived nutrients are concentrated where runoff enters the marine environment in a confined channel. The World Resources Institute has identified 375 hypoxic coastal zones around the world, concentrated in coastal areas in Western Europe, the Eastern and Southern coasts of the US, and East Asia, particularly in Japan. In the ocean, there are frequent red tide algae blooms that kill fish and marine mammals and cause respiratory problems in humans and some domestic animals when the blooms reach close to shore.

In addition to land runoff, atmospheric anthropogenic fixed nitrogen can enter the open ocean. A study in 2008 found that this could account for around one third of the ocean's external (non-recycled) nitrogen supply and up to three per cent of the annual new marine biological production. It has been suggested that accumulating reactive nitrogen in the environment may have consequences as serious as putting carbon dioxide in the atmosphere.

One proposed solution to eutrophication in estuaries is to restore shellfish populations, such as oysters. Oyster reefs remove nitrogen from the water column and filter out suspended solids, subsequently reducing the likelihood or extent of harmful algal blooms or anoxic conditions. Filter feeding activity is considered beneficial to water quality by controlling phytoplankton density and sequestering nutrients, which can be removed from the system through shellfish harvest, buried in the sediments, or lost through denitrification. Foundational work toward the idea of improving marine water quality through shellfish cultivation to was conducted by Odd Lindahl et al., using mussels in Sweden.

Plastic Debris

A mute swan builds a nest using plastic garbage.

Marine debris is mainly discarded human rubbish which floats on, or is suspended in the ocean. Eighty percent of marine debris is plastic – a component that has been rapidly accumulating since the end of World War II. The mass of plastic in the oceans may be as high as 100,000,000 tonnes (98,000,000 long tons; 110,000,000 short tons).

Discarded plastic bags, six pack rings and other forms of plastic waste which finish up in the ocean present dangers to wildlife and fisheries. Aquatic life can be threatened through entanglement, suffocation, and ingestion. Fishing nets, usually made of plastic, can be left or lost in the ocean by fishermen. Known as ghost nets, these entangle fish, dolphins, sea turtles, sharks, dugongs, crocodiles, seabirds, crabs, and other creatures, restricting movement, causing starvation, laceration and infection, and, in those that need to return to the surface to breathe, suffocation.

Remains of an albatross containing ingested flotsam

Many animals that live on or in the sea consume flotsam by mistake, as it often looks similar to their natural prey. Plastic debris, when bulky or tangled, is difficult to pass, and may become permanently lodged in the digestive tracts of these animals. Especially when evolutionary adaptions make it impossible for the likes of turtles to reject plastic bags, which resemble jellyfish when immersed in water, as they have a system in their throat to stop slippery foods from otherwise escaping. Thereby blocking the passage of food and causing death through starvation or infection.

Plastics accumulate because they don't biodegrade in the way many other substances do. They will photodegrade on exposure to the sun, but they do so properly only under dry conditions, and water inhibits this process. In marine environments, photodegraded plastic disintegrates into ever smaller pieces while remaining polymers, even down to the molecular level. When floating plastic particles photodegrade down to zooplankton sizes, jellyfish attempt to consume them, and in this way the plastic enters the ocean food chain. Many of these long-lasting pieces end up in the stomachs of marine birds and animals, including sea turtles, and black-footed albatross.

Plastic debris tends to accumulate at the centre of ocean gyres. In particular, the Great Pacific Garbage Patch has a very high level of plastic particulate suspended in the upper water column. In samples taken in 1999, the mass of plastic exceeded that of zooplankton (the dominant animal life in the area) by a factor of six. Midway Atoll, in common with all the Hawaiian Islands, receives substantial amounts of debris from the garbage patch. Ninety percent plastic, this debris accumulates on the beaches of Midway where it becomes a hazard to the bird population of the island. Midway Atoll is home to two-thirds (1.5 million) of the global population of Laysan albatross. Nearly all of these albatross have plastic in their digestive system and one-third of their chicks die.

Marine debris on Kamilo Beach, Hawaii, washed up from the Great Pacific Garbage Patch

Toxic additives used in the manufacture of plastic materials can leach out into their surroundings when exposed to water. Waterborne hydrophobic pollutants collect and magnify on the surface of plastic debris, thus making plastic far more deadly in the ocean than it would be on land. Hydrophobic contaminants are also known to bioaccumulate in fatty tissues, biomagnifying up the food chain and putting pressure on apex predators. Some plastic additives are known to disrupt the endocrine system when consumed, others can suppress the immune system or decrease reproductive rates. Floating debris can also absorb persistent organic pollutants from seawater, including PCBs, DDT and PAHs. Aside from toxic effects, when ingested some of these are mistaken by the animal brain for estradiol, causing hormone disruption in the affected wildlife.

Toxins

Apart from plastics, there are particular problems with other toxins that do not disintegrate rapidly in the marine environment. Examples of persistent toxins are PCBs, DDT, TBT, pesticides, furans, dioxins, phenols and radioactive waste. Heavy metals are metallic chemical elements that have a relatively high density and are toxic or poisonous at low concentrations. Examples are mercury, lead, nickel, arsenic and cadmium. Such toxins can accumulate in the tissues of many species of aquatic life in a process called bioaccumulation. They are also known to accumulate in benthic environments, such as estuaries and bay muds: a geological record of human activities of the last century.

Specific Examples

- Chinese and Russian industrial pollution such as phenols and heavy metals in the Amur River have devastated fish stocks and damaged its estuary soil.

- Wabamun Lake in Alberta, Canada, once the best whitefish lake in the area, now has unacceptable levels of heavy metals in its sediment and fish.

- Acute and chronic pollution events have been shown to impact southern California kelp forests, though the intensity of the impact seems to depend on both the nature of the contaminants and duration of exposure.

- Due to their high position in the food chain and the subsequent accumulation of heavy metals from their diet, mercury levels can be high in larger species such as bluefin and albacore. As a result, in March 2004 the United States FDA issued guidelines recommending that pregnant women, nursing mothers and children limit their intake of tuna and other types of predatory fish.

- Some shellfish and crabs can survive polluted environments, accumulating heavy metals or toxins in their tissues. For example, mitten crabs have a remarkable ability to survive in highly modified aquatic habitats, including polluted waters. The farming and harvesting of such species needs careful management if they are to be used as a food.

- Surface runoff of pesticides can alter the gender of fish species genetically, transforming male into female fish.

- Heavy metals enter the environment through oil spills – such as the Prestige oil spill on the Galician coast – or from other natural or anthropogenic sources.

- In 2005, the 'Ndrangheta, an Italian mafia syndicate, was accused of sinking at least 30 ships loaded with toxic waste, much of it radioactive. This has led to widespread investigations into radioactive-waste disposal rackets.

- Since the end of World War II, various nations, including the Soviet Union, the United Kingdom, the United States, and Germany, have disposed of chemical weapons in the Baltic Sea, raising concerns of environmental contamination.

Underwater Noise

Marine life can be susceptible to noise or the sound pollution from sources such as passing ships, oil exploration seismic surveys, and naval low-frequency active sonar. Sound travels more rapidly and over larger distances in the sea than in the atmosphere. Marine animals, such as cetaceans, often have weak eyesight, and live in a world largely defined by acoustic information. This applies also to many deeper sea fish, who live in a world of darkness. Between 1950 and 1975, ambient noise at one location in the Pacific Ocean increased by about ten decibels (that is a tenfold increase in intensity).

Noise also makes species communicate louder, which is called the Lombard vocal response. Whale songs are longer when submarine-detectors are on. If creatures don't "speak" loud enough, their voice can be masked by anthropogenic sounds. These unheard voices might be warnings, finding of prey, or preparations of net-bubbling. When one species begins speaking louder, it will mask other species voices, causing the whole ecosystem to eventually speak louder.

According to the oceanographer Sylvia Earle, "Undersea noise pollution is like the death of a thousand cuts. Each sound in itself may not be a matter of critical concern, but taken all together, the noise from shipping, seismic surveys, and military activity is creating a totally different environment than existed even 50 years ago. That high level of noise is bound to have a hard, sweeping impact on life in the sea."

Adaptation and Mitigation

Aerosol can polluting a beach.

Much anthropogenic pollution ends up in the ocean. The 2011 edition of the United Nations Environment Programme Year Book identifies as the main emerging environmental issues the loss to the oceans of massive amounts of phosphorus, "a valuable fertilizer needed to feed a growing global population", and the impact billions of pieces of plastic waste are having globally on the health of marine environments. Bjorn Jennssen (2003) notes in his article, "Anthropogenic pollution may reduce biodiversity and productivity of marine ecosystems, resulting in reduction and depletion of human marine food resources". There are two ways the overall level of this pollution can be mitigated: either the human population is reduced, or a way is found to reduce the ecological footprint left behind by the average human. If the second way is not adopted, then the first way may be imposed as world ecosystems falter.

The second way is for humans, individually, to pollute less. That requires social and political will, together with a shift in awareness so more people respect the environment and are less disposed to abuse it. At an operational level, regulations, and international government participation is needed. It is often very difficult to regulate marine pollution because pollution spreads over international barriers, thus making regulations hard to create as well as enforce.

Without appropriate awareness of marine pollution, the necessary global will to effectively address the issues may prove inadequate. Balanced information on the sources and harmful effects of marine pollution need to become part of general public awareness, and ongoing research is required to fully establish, and keep current, the scope of the issues. As expressed in Daoji and Dag's research, one of the reasons why environmental concern is lacking among the Chinese is because the public awareness is low and therefore should be targeted. Likewise, regulation, based upon such in-depth research should be employed. In California, such regulations have already been put in place to protect Californian coastal waters from agricultural runoff. This includes the California Water Code, as well as several voluntary programs. Similarly, in India, several tactics have been employed that help reduce marine pollution, however, they do not significantly target the problem.

In Chennai, sewage has been dumped further into open waters. Due to the mass of waste being deposited, open-ocean is best for diluting, and dispersing pollutants, thus making them less harmful to marine ecosystems.

References

- Administration, US Department of Commerce, National Oceanic and Atmospheric. "What is the biggest source of pollution in the ocean?". oceanservice.noaa.gov. Retrieved 2015-11-22.

- Hamblin, Jacob Darwin (2008) Poison in the Well: Radioactive Waste in the Oceans at the Dawn of the Nuclear Age. Rutgers University Press. ISBN 978-0-8135-4220-1

- Podsadam, Janice (19 June 2001). "Lost Sea Cargo: Beach Bounty or Junk?". National Geographic News. Retrieved 8 April 2008.

- "Research | AMRF/ORV Alguita Research Projects" Algalita Marine Research Foundation. Macdonald Design. Retrieved 19 May 2009.

- Weiss, Kenneth R. (2 August 2006). "Plague of Plastic Chokes the Seas". Los Angeles Times. Archived from the original on 25 March 2008. Retrieved 1 April 2008.

- Moore, Charles (November 2003). "Across the Pacific Ocean, plastics, plastics, everywhere". Natural History. Archived from the original on 27 September 2007. Retrieved 5 April 2008.

- "Midway's albatross population stable | Hawaii's Newspaper". The Honolulu Advertiser. 17 January 2005. Retrieved 20 May 2012.

- North, W. J.; James, D. E.; Jones, L. G. (1993). "History of kelp beds (Macrocystis) in Orange and San Diego Counties, California". Fourteenth International Seaweed Symposium. p. 277. doi:10.1007/978-94-011-1998-6_33. ISBN 978-94-010-4882-8.

- Warner R (2009) Protecting the oceans beyond national jurisdiction: strengthening the international law framework. Vol. 3 of Legal aspects of sustainable development, Brill, ISBN 978-90-04-17262-3.

Permissions

Index

* 9 7 8 1 6 3 5 4 9 2 0 3 3 *